A New Vision of the Early Universe

Solving the Problems Issuing from the Big Bang Theory

Robert J. Conover

Cover image from Pixabay.com
Illustrations of the photon by Anna Keene
Editing by Donna D. Lawson

ISBN-13: 978-0-9797298-5-0 (Hardback)
ISBN-13: 978-0-9797298-8-1 (Paperback)
ISBN-13: 978-0-9797298-3-6 (EBook)

Library of Congress Control Number: 2022992557

First edition
Printed in the United States of America
January 2023

Science/Space Sciences/Cosmology
Science/Physics/Astrophysics
Philosophy/Metaphysics

InPerspective Publications
PO Box 805
San Luis Obispo, CA 93406

Also by Robert J. Conover

Journey of the Universe

A New Perspective on its Past, Present, and Future Evolution

To my grandchildren:
Aaron, Caitlyn, McKayla, Ava, and Kellan.
Their intelligence, compassion, and determination will make their lives both enjoyable and rewarding, while making the world a better place.

And to my future wife and soulmate, Ellen, whose patience, support, and encouragement are without bounds, and without which this project would not have happened.

Table Of Contents

Introduction

You are about to embark on an amazing construction project, one that will span the entire universe starting at its birth. Together we will create an entirely new vision of what that epoch looked like. Let's begin by examining why this is necessary. What is wrong with our current theory of the early universe?

The short answer is that if we theorize the wrong initial events at its birth, the consequent theories and mathematics supporting those events lead us to conflicts and unanswered questions, which is the position we find ourselves in today. It is time to reexamine those initial events, which serve as the foundation for the story of our early universe.

From the equations of Einstein's General Relativity that predicts the existence of blackholes, it is theorized the universe started from a singularity — a speck of energy infinitely small and dense. The singularity began expanding and particles and forces formed at the Big Bang, creating a very hot, dense, matter state that gradually expanded, cooled, and set the stage for the evolution of our universe as we know it today.[1]

The Standard Model of Particle Physics posits that particles and forces popped into existence within the first second of the Big Bang. A mathematical structure has been created to support that theory, but it is not otherwise well supported and is difficult to believe. Complex systems found in nature such as those don't suddenly *pop* into existence, they evolve.

That Hot Big Bang Theory is a part of the Standard Model of Cosmology, and since it and the Standard Model of Particle Physics each borrow heavily from each other in how they view the early universe, they will henceforth be referred to together simply as the Standard Model. Evidence for the Hot Big Bang Theory is supported by the following observational evidence:[2]

- The universe quickly expanded and cooled.
- Atoms formed allowing photons to scatter producing the Cosmic Microwave Background.
- Hydrogen and helium gases formed in the ratios predicted.
- Galaxies evolved in the structure of a cosmic web that is predicted mathematically and confirmed by computer simulation.

True enough, all that observational evidence supports the notion the universe began as a tiny speck of expanding energy from which particles and forces were created and evolved. But that same evidence could support a variety of different beginnings, including the beginning espoused herein. As we shall examine in Chapter One, the evidence for the timing and creation of supermassive blackholes, the creation of the cosmic web (*think formation of the galaxies*), and the reason the universe is so homogeneous, all suggest a beginning different than that of the Hot Big Bang Theory.

Starting from a singularity might make sense mathematically, but it sends us down a path having far too many unanswered questions. As we know, no matter how elegant the math, it can be plagued with wrong turns. As Physicist Michio Kaku reminds us, beauty in mathematics can lead you astray.[3]

We need a new story for the beginning of our universe, one premised on natural evolution, rather than unsupported events and mathematics. Physicist Harry Cliff tells us, 'This is a moment to reexamine our assumptions and look at old problems from a different angle. More than anything, it is time to put our grand ideas and preconceptions to one side and listen carefully to what nature is saying.'[4] That is exactly the specifications we will follow while reconstructing our early universe. We will listen to nature by carefully analyzing our observations, leaving as little to fantasy as possible.

Breakthroughs come when scientists discard a cherished assumption.[5] In keeping with that notion, readers are asked for the duration of this book to temporarily suspend any assumptions they may have regarding how the universe began, and specifically any belief in the Hot Big Bang Theory. Please view the early universe temporarily with an open mind and a blank slate.

Our new vision of the early universe will start with a high-density speck of energy and will have a Big Bang moment, but without any preconceived notions as to what results from those conditions. We will build an entirely new early universe, one that begins differently but naturally evolves into exactly the universe we see today. While taking notice of and giving due respect to the Standard Models, we will proceed to build our early universe without the mathematical structure on which those models depend.

For some readers, exploring a new theory without a mathematical basis will seem like a waste of time. Those readers are reminded that some of the greatest accomplishments in physics, such as Newton's theory of gravity, Maxwell's theory of electromagnetism, and Einstein's theory of relativity, all started from a theoretical basis before the complete mathematical structure was finalized. Even some of Quantum Mechanics came from pure theoretical reasoning.6 To make our construction project a solid theory we will need a good mathematical structure, but for now we are simply exploring to see whether our construction venture produces something worthy of a mathematical structure.

Physicist Lisa Randall informs us '...ultimately critical thinking is the only reliable way to answer questions about the makeup of the universe.'[7] We will employ that critical thinking as much as possible, recording our observations and making reasonable deductions. Our first three deductions in Chapter One are noted as *Tentative Deductions* because, though quite reasonable, their support is minimal, and they represent a notable departure from current theory. Our fourth, fifth, and sixth deductions, however, are well supported with a significant amount of observational evidence that validates the first three deductions.

Theoretical and practical progress results from a single human activity — good explanations.8 Armed with only an open mind and a blank slate upon which to record our observations and deductions, we shall attempt good explanations. We shall also stay in keeping with the scientific goals of creating a theory that is simple, clear, natural, and elegant. We now boldly go where no one has gone before, to the dawn of the universe armed with only a blank slate and no expectation as to what we will find there. Our journey begins!

Chapter One

Building a Foundation

Initial Observations

Our mission is to build a new picture of our early universe from scratch using only our keen observations and deductions. Let's start by observing the gross aspects of the universe we see all around us. We observe galaxies, blackholes, stars, planets, moons, and gases, all of which we simply characterize as matter.

Over the past century, scientists have been drilling into matter exposing its layers. An early examination confirmed the long-held belief that it was comprised of tiny bits called atoms. Drilling deeper they discovered atoms comprise a nucleus orbited by different configurations of electrons. Drilling into the nucleus revealed it to be made up of two particles, labeled protons and neutrons. Going deeper they found that both are in turn made up of two different particles called quarks. We do not yet have the technology to drill deeper into matter than the quark. The known particles within the atom are referred to as subatomic particles, hereinafter called *Sub-A's*.

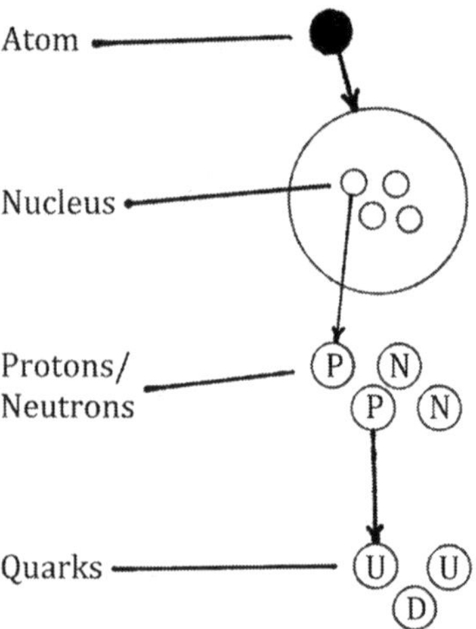

Figure 1.1 The drilling into matter from the atom down to quarks.

Matter is important to us not only because we are made of it, but because it tells us something very important about the universe. According to Einstein's famous equation $E=MC^2$, energy (E) is equal to mass (M) times the speed of light (C) squared. It means energy and mass are equivalent; everything that is mass is made from this stuff called energy. We know what mass looks like, but what is energy?

Energy

Scientists speak of energy in one of its several manifestations such as radiant energy or nuclear energy, but in this case, we are referring to the form of energy from which everything is made — *raw energy*, often called pure energy. For clarity, and to separate it from other forms of energy we may discuss later, pure energy will be called *Penergy*. So, Penergy is just another name for raw energy, which is the energy from which everything we observe in the universe is made.[1]

Penergy is difficult to define because there is nothing else like it. It's not a solid, liquid, or gas, but it has the capacity to make itself into things we can refer to as such. It is very dynamic, meaning it has the capacity to *be* many things (*blackholes and particles*); to make things *happen* (*evolution and forces*); to *bring about* non-tangible things (*consciousness and intelligence*); and to possess even more capabilities we are only now beginning to realize. The reader will gain a better appreciation for what Penergy is as its many dynamics and characteristics are examined in due course.

Penergy can appear in several *states* or conditions, which will be discussed shortly. Initially it is referred to as being a cloud, but that is only to give the reader a visual image. For the moment, simply think of Penergy as a unique, fluid-like substance, and the only real substance in our entire universe appearing in many manifestations. We will make an in-depth examination of Penergy and how it relates to other forms of energy in Chapter Eight.

We have observed that the entire visible universe is made of matter, and that matter is made of Penergy. To better understand this landscape to our universe we might next ask, *where did the Penergy come from?*

The Origin of Penergy

The short answer is, we don't know where the Penergy came from. That question resolves down to a philosophical issue. We can only make reasonable speculations on the Penergy's initial size and density.

We have observed evidence indicating galaxies in our universe are moving away from each other. From this we can infer the universe is expanding, not just at its edges like a balloon, but from within, like a loaf of raisin bread in the oven expands with all its raisins moving away from each other.[2] In this sense the galaxies aren't really moving on their own. Their only movement is incidental to the space between them expanding, which occurs in all three dimensions, so we view this as the galaxies moving away from each other.

If the entire universe is expanding, it means that sometime in the past it was much smaller. Scientists have wound the clock back imagining a universe that was smaller and smaller. They ultimately calculated the universe to be 13.8 billion years old and have theorized that at its birth it was just a speck of very condensed Penergy.

Astronomer Philip Ball portrays it as the entire universe being smaller than an atom.[3] Astrophysicist Neil deGrasse Tyson tells us that at the beginning of time, all space, matter, and energy in the universe fit within a pinhead.[4] Most authors on the subject say about the same thing, so let's just agree the initial speck of Penergy was very small. Some say that it was *infinitely* small. But what does that mean?

The Singularity Problem

Within weeks of Einstein publishing his theory of General Relativity in 1915, Physicist Karl Schwarzschild used Einstein's concept to demonstrate that an object in space with sufficiently strong gravity could create something we now call a blackhole.[5] Soon thereafter, Russian Mathematician Alexander Friedmann found solutions to Einstein's equations describing universes evolving from a singular state of infinite density — what is now called a *singularity*.[6]

In the 1970's, Physicist Stephen Hawking and Mathematician Roger Penrose wrote a theorem concluding that the energy kernel at the start of an expanding universe such as ours, and at the core of all blackholes, is a singularity.[7] Hawking later wrote that he changed his mind and that there was no singularity at the start of the universe.[8] Still, the idea of the universe starting from a singularity with infinite energy and density persists. Astrophysicist Carolyn Devereux tells us in her recent book, *Cosmological Clues,* that scientists do not know how else to describe the start of the universe and they put up with the singularity idea until something better comes along.[9]

The word *infinite* here has its usual meaning — something without limit that goes on forever. We can imagine something infinite, like a number line to which one can always add one more number, but all such examples are simply human-made concepts, not reality. Infinity is an interesting idea and may even be useful in mathematical calculations, but to believe that anything in nature goes on forever is a challenge. We have not observed a singularity or anything else in nature that is infinite.[10] Infinite density and anything else one might deem infinite is only theoretical.

Infinities create their own challenges for scientists. As Astronomer Chris Impey tells us, mathematics has many ways to manipulate and deal with infinities, but in physics infinity is a big problem.[11] Physicist Michio Kaku tells us that to a physicist, infinity is just a sign that the equations are not working; that the physicists don't understand what is happening.[12] Einstein believed the presence of singularities was a sign of imperfect physical understanding and that it made no sense for an object to have zero size and infinite mass density.[13]

The concept of a singularity seems to present a conflict in ideas. If something has *infinite* density, how is it ever going to change short of a God-like intervention. If its density is infinite, it sounds pretty hard, unbreakable, and unchangeable. If its density is infinite, it can't form anything else. If it becomes something else, its density must not have been infinite.

Anything infinite is an abstraction; purely a mathematical object.[14] As we know, mathematics can be wonderful, but its equations don't always represent reality. As Michio Kaku again tells us, we cannot blindly accept the math of Einstein's theory since his equations predict the center of a blackhole or the beginning of time as infinite, which makes no sense.[15] For all these reasons we will keep things simple and not infuse any infinities or singularities into our work here. Our blueprint for constructing our new early universe will call for a non-singularity; a simple *speck* of Penergy will do.

We don't need to resolve how dense the speck of Penergy was but agree it was very dense; a condition I like to call the NIB state — **N**ear **I**nfinite **B**ut (not quite). For reasons unknown, the speck of very-dense Penergy changed and the point in time that change occurred is known as the Big Bang. We don't know how the speck of Penergy originated, how long it existed prior to becoming our universe, or what caused it to change. Any speculations on those questions would only detour us down a philosophical path away from our goal so we won't go there.

We know little about the speck of Penergy beyond the fact that apparently our entire observable universe is made from it. Anything else we learn about it will have to be derived from our observations — measured, deduced (*verbally or mathematically*), or surmised. The next observation we might make is that the Penergy somehow turned itself into matter. That observation of course brings up the questions of *when, how,* and *from what density* of Penergy? Let's see what observations we can make that may answer these important questions.

The Penergy Density Question

The Standard Model posits that in the first millionth of a second following the Big Bang all the basic elements and forces were formed, creating the Sub-A's we know today.[16] In all due respect for this Hot Big Bang Theory, that story is difficult to believe. In fact, further observations about Penergy that will be related throughout this chapter suggest the universe started quite differently.

If Penergy started out as a high-density speck and somehow changed to a mass of comparatively light-density particles, it is more reasonable that it would have accomplished that feat over time in a series of steps, rather than in a single, sudden, explosive episode. Going from one extreme of a near-infinite density speck, to another extreme of various, complex, light-density particles, inside an explosion lasting a fraction of a second might be possible, but is unsupported, unreasonable, and unnecessary.

We do not know the dissipation phases of Penergy at the start of a universe any more than we understand the condensing phases of matter at the end of its life inside a blackhole. The primordial universe at the Big Bang, like the blackhole, is a mystery beyond the reach of science's physical models.[17] Based on the phase transitions (*solid→liquid→gas*) we know matter to be capable of, it is reasonable to believe that multiple, specific, and dynamic density phases take place both inside a blackhole and at the unfolding of a universe.

The apparent change in Penergy density from very dense to relatively light suggests that Penergy is capable of existing in a range of densities. Why would something so important and dynamic as Penergy, the substance of our entire universe, be limited to existing in only two density states? And two densities that are not close to or even related to each other? Penergy having a density makes a lot of sense. Penergy being confined to only two density-states makes no sense. Given these arguments (*and more to follow*), we can make our first tentative deduction.

> Tentative Deduction #1: If Penergy initially existed in a near infinite-density state as believed, and we know it to exist in a comparatively light-density state as particles, we can infer that Penergy can exist in a range of densities. Additional support for this deduction will be revealed in subsequent observations and deductions.

When Was the First Matter Created?

The standard Model's story of complex forces and particles inexplicably popping into existence within the first second of the big bang stretches credulity. Protons and neutrons are complex, believed to be made up of two different types of quarks and a sea of eight kinds of gluons. Why would these complex particles form inside the first second of the Big Bang with no purpose, cause, blueprint, or evolutionary basis?

It makes much more sense for particles, especially complex-composite particles such as protons and neutrons comprised of many different smaller particles, to have come from something simpler. Again, complex systems found in nature such as those don't suddenly *pop* into existence, they *evolve*. How that could have happened will be addressed in the following chapters. For now, the point is that a high-density state of Penergy suddenly changing into a comparatively light-density state of complex particles is not particularly believable, especially since there are more rational alternatives. This leads us to our second tentative deduction.

> Tentative Deduction #2: If Penergy initially existed in a near infinite-density state, there is no compelling reason to believe it suddenly changed into a mass of many different types of light-density, complex particles within the first second of the Big Bang, as theorized. In short, our subatomic particles were not likely created within the first second of the Big Bang. Additional support for this deduction will be revealed in subsequent observations and deductions.

If the near-infinite speck of Penergy did not convert to particles instantly at the Big Bang, it suggests the Penergy had to await a lesser density for light-density particles to form. For the Penergy density to lessen, the volume of the universe would have to grow. Our baby universe of Penergy would have to expand, as we know it does. But what was, and is, causing the expansion?

Cosmic Expansion (*What is Dark Energy?*)

Since Astronomer Edwin Hubble in 1929 observed that galaxies were moving away from each other concluding the universe was expanding, scientists have speculated on what is causing that expansion. The consensus is that it must be in a form of energy, but they know of no existing energy that could do that or how it can even exist.[18] The mysterious energy has been dubbed *Dark Energy*.

The Standard Model says everything was created from the speck of Penergy within the first second of the Big Bang, with most of the energy in the form of *light* and the rest in the form of fundamental particles, mainly protons, neutrons, electrons, and positrons.[19] The Standard Model seems to assume that in the creation of those particles of light and matter, all the Penergy in the original speck was consumed. We have already concluded that the notion of complex particles popping into existence within the first second is unreasonable, so the later assumption that all the Penergy was consumed within that second is also unreasonable.

Let's see what more we can learn from the two premises we started with; the universe started out as a speck of very-dense Penergy and for reasons unknown began to change by expanding. We can infer that Penergy when at some density less than its very-high density NIB-state, naturally expands. As the universe expanded and the

Penergy thinned, we know it eventually reached a density that allowed particles to come into existence. There is no reason to believe all the Penergy was consumed even in that particle creation epoch. That would mean a thin density volume of Penergy might still exist, pervading all of space.

That prospect is not only real but obvious since according to quantum theory virtual particles (*think particles that quickly pop in and out of existence*) can be momentarily created by borrowing energy from the vacuum [*of space*].[20] What other energy could those virtual particles borrow from if not the same Penergy from which all real particles and everything else in the universe is made? For virtual particles made of Penergy to pop into existence, the Penergy must be present in the vacuum of space. According to Physicist Lisa Randall, when a proton and anti-proton annihilate, they turn back into pure energy.[21] That *pure energy* is, of course, Penergy, comprising the space into which those particles disappear.

Mario Livio points out in his book, *The Accelerating Universe,* that particle-antiparticle pairs can also borrow energy from the vacuum, and those particles immediately annihilate themselves back into the vacuum. He characterizes the vacuum as bubbling with such virtual pairs.[22] Again, if known particle pairs are being momentarily produced out of the vacuum, that vacuum must contain the requisite energy to produce those pairs.

The creation of particle pairs implies that the energy of the so-called vacuum of space is the same energy from which the first real particles were created, the energy from the original speck — Penergy. We can conclude from all of this that the vacuum of space is not actually a vacuum but is filled with an energy capable of producing virtual particles and particle pairs, and the only energy capable of doing that as far as we know is the same energy from which all other particles are made — Penergy.

If all the Penergy was not consumed creating particles in the early universe, it would leave the residual Penergy growing even thinner as the universe expanded, until the Penergy was too thin to create new particles. Though thin, apparently fluctuations in the existing Penergy density still allows particles to momentarily come into existence.[23] This suggests the current Penergy density is just below the particle-creation density threshold but is dense enough for fluctuations in that density to momentarily create particles.

This remaining very-thin Penergy could now simply serve as our interspatial medium and be the source of energy that is continuing to expand the universe. Dark Energy could be nothing more than a very-thin-density version of the Penergy that started the universe. The same Penergy from which everything is made and from which virtual particles and pairs borrow, and into which they disappear.

We know that the interspatial medium consists of something, because scientists report evidence that the spin of the earth [*and all other large bodies*] is causing *frame dragging,* meaning the earth's spin is dragging the adjacent spacetime with it.[24] Possibly the actual substance being dragged is the thin Penergy, a.k.a., dark energy.

Further evidence our interspatial medium is a fluid-like substance and not a vacuum comes from evidence of gravitational waves, which were predicted from Einstein's theory of General Relativity. Gravitational waves are disturbances in space caused by cataclysmic events such as the crashing together of neutron stars or black holes. Those disturbances create waves that propagate across space. Scientists have built the very sensitive equipment necessary for their detection and in 2016 reported that the first gravitational waves were detected.[25]

Gravitational waves stretch and distort everything along their path. A circle becomes an ellipse, and a square becomes a rectangle.[26] According to Jorge Cham and Daniel Whiteson, this kind of behavior can only happen if space has a certain physical nature to it.[27] There must be something doing the waving.

At one time scientists thought space contained a substance they called *aether*, but the presence of such an aether was disproven in the famous Michelson-Morley experiment. The presence of the Penergy may be difficult to prove, as well. It is not a physical, aether-like substance made from particles that would have mass or other particle characteristics that we can detect. Pure energy may not be directly detectable, but we might be able to infer its presence from its relationship to particles and virtual particles, as discussed above.

Einstein's theory of General Relativity taught us that gravity is the distortion of spacetime, which is often referred to as a fabric. That notion implies that the fabric of spacetime is something that can be curved and distorted since that is the only reasonable way it could influence the path of passing matter. Astrophysicist Paul Sutter says that General Relativity taught us that spacetime itself is a dynamic, living, breathing, physical object.[28]

We also know that space can be made into waves and can be expanded. All these characteristics make more sense if they apply to the existence of *something*. That something may be difficult to detect because it is not like anything else we encounter, but there is little doubt the vacuum of space is no vacuum at all but is filled with something. The something that makes most sense is a thin-density version of the original Penergy from which the universe began its expansion. Based on the above observations we can make our third tentative deduction.

Tentative Deduction #3: If the original Penergy speck was capable of multiple densities and was not immediately consumed in particle creation, some thin-density Penergy may still exist today comprising our interspatial medium. If everything in the universe is made of Penergy, then it stands to reason that the now very thin-density Penergy is the Dark Energy that has been expanding the universe since its inception. Additional support for this deduction will be revealed in subsequent observations and deductions.

We now have Penergy existing initially in a very-high *NIB-density* state, in a light *particle-density* state, and in a very-light *dark-energy-density* state. As we shall discuss shortly there are more density states in our cosmic history. Obviously Penergy density has been very important in the evolution of the universe so let's take a closer look at the idea of *density*.

Penergy Density Scale

The notion of density derives from the relationship between substance and volume. *Substance* can be most anything, solid, liquid, or gas. *Volume* is a defined area. Density is the degree of compactness of a substance in a given area. In a measured amount of substance, the smaller the volume we place it in the greater its compaction or density. If we spread that substance over a greater volume, we have reduced or thinned its density. Think of high density as lots of stuff packed into a small volume. Density is mathematically expressed in various ways depending on the substance involved, but generally it is Density=Substance/Volume.

Using the above definition, as the universe expanded in volume, the Penergy substance would have become thinner in density. With regards to Penergy however, the thinning was not in direct proportion to its expansion. Scientists tell us that the energy expanding the universe is *growing* as it expands.[29] That means, as it expands from within (*like raisin bread*), more Penergy is produced, thus it's not like a fixed amount growing thinner with more volume. With Penergy, the thinning density rate is much slower than its expansion rate. As the Penergy thins, more Penergy is pulled out of the existing Penergy, and apparently the thinner it gets the more pressure there is to extract more.

This seems to be a violation of the first law of thermodynamics, which states that energy cannot be created or destroyed. In this case it is not that new energy is being created out of nothing, it is simply the dynamics of how Penergy thins by drawing more out of itself. It's not the expanding universe that is causing the thinning, it is apparently the thinning of the Penergy pushing out to expand and create whatever volume it needs. This is largely theoretical and for our project we need not pursue it further.

We could make up a density scale to track the phase transitions the Penergy may have gone through. If the near-infinite, NIB-state density was given a value of 1.0, we could break the scale down into increments of .1, giving it .9, .8, .7 densities, etc.

If the 1.0 represents a near-infinite density (*near solid state*) of Penergy, then the .9 and .8 densities that came into existence as it thinned would still be pretty dense, perhaps too dense for anything significant to have happened. At .7 density, however, the Penergy might be thin enough to allow things to happen with which we can identify. During the thinning from 1.0 to .7 density the universe could have grown substantially in size, especially if the expansion rate was in some way proportional to the energy density as some physicists have suggested.[30]

Some readers may think the Tentative Deductions we've made so far are borderline and wonder if there is additional observational evidence to support those deductions. Yes, there is. It is strong evidence, and it comes from some of the oldest creations in the universe — supermassive blackholes. Their existence, as well as other observations and deductions we will make later, all lend a great deal of validity to our first three Tentative Deductions.

Blackholes

Most of the universe is empty space except for various concentrations of huge galaxies, each made up of a supermassive blackhole surrounded by a plane of billions of stars. What can we learn from this observation? First let's look at the elephant in the picture — the blackhole.

A blackhole is a region of space where gravity is so strong that nothing can escape its boundary called the *event horizon.* Blackholes were predicted from Einstein's field equations, for many years believed to possibly exist, and finally in the 1990's they were confirmed to exist.

The blackhole creation theory is that a mass with sufficient compaction (*density*) could have enough gravitational influence to turn it into a blackhole. To get an idea of how much compaction we are talking about, for the planet earth to become a blackhole it would have to have its entire mass compressed down to a sphere the size a little smaller than a ping-pong ball.[31]

A star several times bigger than our sun when it reaches the end of its life is theorized to implode into a much smaller, denser mass that goes through various stages, condensing further and further, eventually becoming a blackhole. It is theorized that this stellar collapse into a blackhole is the normal life cycle of stars that are eight to one hundred times bigger than our sun.[32] There are relatively few that size however, less than one per cent of the observable stars.[33]

Blackholes make it apparent that Penergy has the capacity to both expand and condense. It only condenses when a glob of it is in a very dense state and has the opportunity to spin, which gives it a gravitational influence. This idea is consistent with our project so far.

If the initial Penergy speck started out in a state of very high density and suddenly ruptured and spewed raw energy, it would likely have attempted to regain its spin rate and strong gravitational influence. But if the rupture was too great to overcome, it would have continued spewing and spreading. While the interior portions of the Penergy cloud would have been working to condense, the much thinner-density outer edges of the cloud would have been working to expand. Obviously, the expansion dynamic won the contest, and our nascent universe was born.

Blackholes are not yet well understood. According to general relativity, every blackhole contains a singularity, yet inside the blackhole Einstein's field equations become non-sensical.[34] Perhaps the inside of a blackhole could be better understood if the calculations were based on its NIB-state rather than a singularity, and its mass based on the density of its Penergy rather than on its apparent gravitational influence.

Afterall, it is theorized that the particle mass inside a blackhole is being condensed back to Penergy and the Penergy itself is being condensed further and further, until it is presumably back to its NIB-state. Consequently, a blackhole does not actually need a mass of particles to come into existence or sustain itself. The mass of a blackhole seems more about condensed Penergy than about particle matter.

Supermassive Blackholes

Supermassive blackholes (*SMBHs*) are millions, even billions, times bigger than our sun. Scientists have not yet agreed on a sound theory for SMBH creation. Basically, they are not supposed to exist. Scientists are hamstrung trying to develop a theory for SMBH creation *after* atoms and stars were created in keeping with the Standard Model's version of the Big Bang.

Trying to base a SMBH creation theory on the Hot Big Bang platform doesn't seem to work. Blackhole growth simply by accretion (*eating surrounding stars and material*) is rather slow. Chris Impey tells us that in a year, a blackhole can grow by no more than a third of the Moon's mass.[35] At that rate it would take far too long to grow to supermassive size. Under the Hot Big Bang Theory, the only way SMBHs could have grown to millions of solar masses is through blackhole mergers. Let's examine that possibility.

Arguments For and Against SMBH Creation by Mergers

The Hubble Telescope has produced deep-space images showing galaxies that have been calculated to have formed during the first billion years after the big bang.[36] This does not allow enough time for their blackholes to have grown to SMBH size simply from isolated mergers.[37] The merger theory might work however, if the merging blackholes were uncharacteristically large.

It is reported that Astronomer Martin Rees developed a theory that larger blackholes could have started from seed blackholes created by the collapse of thousands of stars, and then those larger blackholes could have merged to create SMBHs.[38] This is a rather complex formula for the creation of a very large blackhole that may require a seed mass of as much as 10,000 solar masses.[39] That complex scenario may have happened somewhere in the universe, but that scenario doesn't seem plausible for the creation of every SMBH at the heart of billions of galaxies.

Computer simulations have been tweaked to the point of allowing very large blackholes to have formed in the early universe. But all those simulations eventually require mergers to take place for SMBHs to be created within two to three billion years after the big bang. Mergers at that age of the universe would require whole galaxies to have merged over and over to grow to a SMBH size. Again, that is an unlikely prospect for creating a SMBH for the billions of galaxies in the universe.

Observational evidence regarding spiral galaxies with SMBHs suggests their formation could not have happened through a frenzy of galaxy mergers. Spiral galaxies have a center mass with arms growing out around it, as shown in Figure 1.2a.

It is believed that when spiral galaxies collide and their blackholes merge, the process tends to scatter the spiral-disk shape.[40] With multiple mergers, the consolidation would create a diffused elliptical galaxy, as shown in Figure 1.2b.

Figure 1.2a Spiral galaxy

Figure 1.2b Elliptical galaxy

Seventy percent of the galaxies in our universe are spiral in shape.[41] The fact that we can observe so many pristine, spiral galaxies containing a supermassive blackhole suggests those galaxies have not suffered the turmoil of multiple mergers and apparently acquired their supermassive blackhole by other means. The remaining thirty percent of galaxies are elliptical and generally house a much bigger supermassive blackhole *billions* of times bigger than our sun. They look like a near spherical cloud, and due to their size and appearance they indeed seem to be the remnants of galaxy mergers.[42]

If SMBHs were not formed by mergers, how were they created? Given our Tentative Deduction#3, that the Penergy starting the universe constantly thinned as the universe expanded, there is a very plausible explanation for how these supermassive blackholes came about. The short answer is that they were formed out of Penergy when it was quite dense in the early universe. Let's examine how that may have happened.

Alternative Theory for SMBH Creation

One thing we have observed about blackholes is that they spin, which seems to be a natural dynamic of Penergy. There is plenty of evidence for forms of Penergy naturally spinning. We know that particles spin, planets spin, stars spin, galaxies spin, blackholes spin, and in the case of stars, they start the process naturally on their own accord. Practically everything in the universe spins or is made of something spinning. It is what Penergy naturally does. With this notion in mind, how would that effect SMBH creation?

If stars can start the process of their own creation by swirling the substance from which they are made, there is no reason to doubt that Penergy can do the same. As the universe expanded and Penergy thinned, at perhaps .7 density the Penergy would have begun to create huge swirls. The developing swirls would be large and quite heavy so the movement would initially be rather slow. But eventually, just like star formation, as the circular motion gradually sped up it would cause the Penergy to pull in toward the center of the swirls, leaving thinner-density areas at their outer edges.

At those outer edges, now at say a .6 density, many more swirls would have begun, drawing in more Penergy, leaving their outer edges less dense. Those outer edges, now at say .5 density, would have created even more swirls, again with their outer edges thinning, setting off a *cascade of swirls* of lesser and lesser densities (*see Figure 1.3*).

Soon there would have been thousands, millions, and billions of swirls of different sizes, causing all kinds of anomalies in their distribution. As the swirls became better defined, they would have begun to create a weak gravitational influence that would have slowed the Penergy expansion near them in proportion to their density, causing even more anomalies in their distribution.

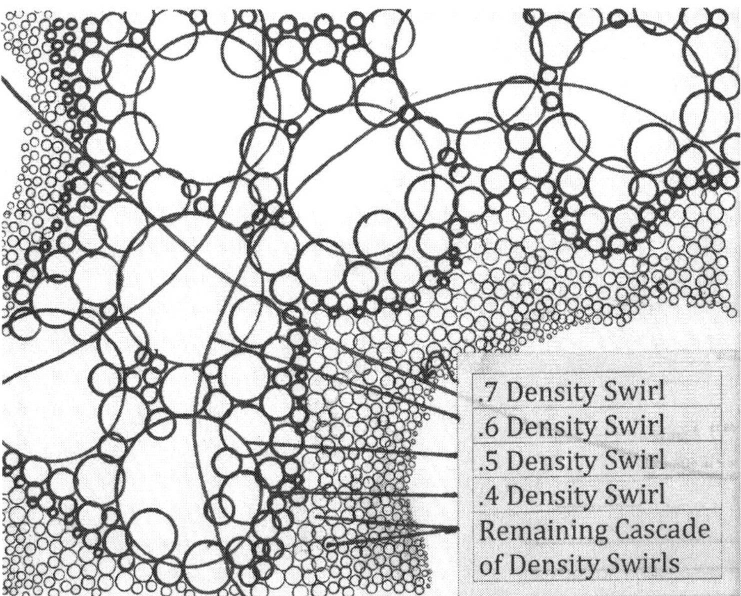

Figure 1.3 The Cascade Effect, showing how each swirl (*depicted by circles*) drew in Penergy, causing its outer edge to thin in density, allowing new, smaller swirls to begin. This effect happened over and over, cascading through smaller and smaller swirls with diminishing densities.

Over time, the bigger swirls would continue condensing, gaining angular momentum with a growing gravitational influence. Eventually the swirls with sufficient density and angular momentum to sustain themselves would turn into *pure energy blackholes*. These blackholes, many million times the size of our sun, will eventually be at the core of billions of galaxies making up our universe. The fluctuations, contractions, territorial battles for Penergy, and the *cascade effect*, would cause the resulting galaxies to form in clusters, groups, walls, filaments, and other peculiar distributions that would ultimately shape our universe.

Pure-energy blackholes are not yet recognized by the Standard Model but may very well exist. An article in *Quanta Magazine* in December 2017, entitled *Earliest Blackhole*

Gives Rare Glimpse of Ancient Universe, discussed the discovery of a very old blackhole with its age placed at 690 million years after the Big Bang and its size equal to 780 million times the mass of our sun. The researchers' calculations discovered that if the blackhole started out as a collapsed star like a blackhole is normally formed, it would not have had time to grow to its enormous size. They further calculated that if it came into existence soon after the Big Bang, it would have had to start out about 1000 times the mass of our sun.

The researchers speculated that perhaps hydrogen/helium gases in the early universe condensed into a blackhole, though admittedly this is inconsistent with scientists' current view of the early universe. Since then, scientists have found galactic blackholes measuring several billion times the size of our sun.[43] Given these reports and the deductions we've made so far, the creation of early, pure energy blackholes seems quite plausible.

Evidence for the Cascade Effect and Pure-Energy Blackhole Creation

The premise here is that the pure-energy supermassive blackholes at the center of all galaxies were structured by the Cascade Effect described above, and their creation was well underway when the elementary particles and stars evolved. Let's examine the observational evidence supporting this premise.

The Number and Size of Blackholes: SMBHs are known to exist in many sizes, having been measured in hundreds of thousand, millions, and billions of solar masses. Nature makes blackholes ranging over a factor of a billion in mass![44] The smallest SMBH recorded is fifty-thousand solar masses.[45] The vast number of sizes of SMBHs, all far too big to have been created from solar collapse, suggests that the theory of a cascade of blackholes in the early universe is consistent with observations.

Galactic Distribution: Galaxies are not randomly scattered across the universe but are preferentially found in super clusters, some containing thousands of galaxies. Within those super clusters are smaller clusters, different sized groups, walls, sheets, and filaments, with very few isolated galaxies.[46] Within larger individual galaxies are smaller galaxies of various sizes. According to Astronomers Luke Barnes and Geraint Lewis in their recent book, *The Cosmic Revolutionary's Handbook*, many *dwarf galaxies* were recently found in a plane rotating around our Milky Way galaxy. Dwarf galaxies have also been found in a plane rotating around the nearby Andromeda galaxy.[47]

These observations and discoveries support the idea of a *Cascade Effect*. The cascading swirls left galaxies embedded in clusters and groups down to the level of dwarf galaxies appearing inside larger individual galaxies. The Cascade Effect should have left the smaller swirls more prevalent. This bares out. It is reported that dwarf galaxies are the most common galaxy type in the universe.[48]

The Cascade Effect should have also left the smallest galaxies appearing within the area of the voids. This is exactly what is observed. Astrophysicist Paul Sutter reports that if you zoom in on a portion of the cosmic web and look at the voids you will see dim dwarf galaxies. He goes on to say that upon a close examination of those voids, one sees a faint echo of the cosmic web. The cosmic web can be seen as a series of nested cosmic webs, each level dimmer and smaller than the last.[49] That description sounds very much like observational evidence for the Cascade Effect.

Galactic Movement: Paul Sutter also tells us that the galaxies comprising the cosmic web are moving. The pattern is not static. "Galaxies are buzzing around in orbits with the cluster."[50] This observation is consistent with the Cascade Effect. Many of the galaxies were created while their Penergy was in a swirling motion being dragged around by a larger swirl. There is no reason to believe those SMBHs that eventually formed would have ceased that circular motion, which could easily be interpreted today as *galaxies buzzing around in orbits.*

Galactic Shape: It is observed that both galaxies and solar systems are generally found in the shape of a disk. This may be because they both form in a very similar manner.

Stars form from a high concentration of gas and dust collecting, rotating, and drawing in material until the ball of gas flattens into a disk. Once the star's nuclear fusion begins, at the outer edge of the star's disk the band of residual matter coalesces into planets circling in the same plane creating a solar system.[51]

By the time the cascade drew down to the smallest swirls, the large swirls had picked up sufficient angular momentum to cause the adjacent Penergy and matter to flatten into a disk, ultimately creating the vast number of spiral galaxies we see today. As newly created particles combined into atoms and gas, the atoms and gas swirled into stars.

Once the now SMBH gained sufficient angular momentum it would have pulled in the surrounding gas, matter, and stars, encircling them in the same plane as the original disk. In summary, just as planets are formed in a plane incidental to star formation, stars are formed in a plane incidental to SMBH formation. This also implies that the SMBH had started forming well before the atoms and stars were created.

Relationship between Galactic Stars and SMBH: Every galaxy has a SMBH at its center, and the mass of the blackhole is tightly coupled to the mass of the stars in that galaxy.[52] This observation also suggests the blackhole was already developing when particles and stars evolved, and once established, the blackhole drew in just the correct amount of matter that it could support with its gravitational field. As Chris Impey puts it, "It was as if the stars knew what galaxy they were in."[53]

Matching our Cosmic Structure — the Cosmic Web: As observed, galaxies are not spread uniformly across the universe but are gathered in clusters, groups, and filaments. The overall arrangement has a honeycomb appearance and is commonly referred to as the Cosmic Web. It is described by Wallace H. Tucker in his book, *Chandra's Cosmos,* as 'massive galaxy clusters as nodes that are interconnected through an intricate web of filaments and sheets of tenuous gas and galaxies, with nearly empty regions called voids taking up most of the volume'.[54]

That description is depicted in Figure 1.4 below. A reconstruction of the cosmic web formation has been computer simulated resulting in a complex theory of energy movements having little mention of the contribution of blackholes. Let's examine an alternative theory that accounts for the creation of the cosmic web in a more natural and less complicated way.

Figure 1.3 illustrates how every large .7 density swirl would create smaller .6 density swirls at their edges, that would in turn create .5 density swirls at their edges, creating .4 density swirls at their edges, cascading down to very tiny swirls. This is called the *Cascade Effect.* The larger swirls would eventually become supermassive blackholes. The intermediate and smaller sized swirls, say at .3 density and below, being too weak to sustain themselves would eventually disappear (*discussed in detail below*). Their disappearance would leave very large voids that would eventually make up ninety percent of space.[55]

The intersection of the larger swirls would settle into super clusters and clusters of SMBHs. The voids would be bordered by groups and filaments of galaxies previously mentioned. Within the clusters, the gravitational influence of the galaxies would slow the expansion between them, while the voids would expand more rapidly creating even larger voids.[56] As Physicist Charles Seife tells us, 'We live in a Swiss cheese universe.'[57] The Cascade Effect is very consistent with the creation of the Cosmic Web of galaxies we see today, as illustrated in the artist's depiction in Figure 1.4.

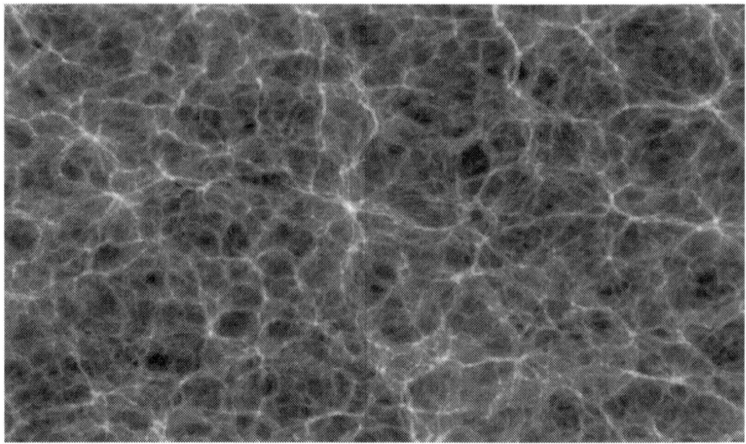

Figure 1.4 Depiction of the cosmic web.

Credit: Virgo – Millennium Simulation Project/Springel et al. (2005)/Max Planck Institute for Astrophysics.

The cosmic web matches exactly the kind of galactic distribution we would anticipate if the galaxies started out as large swirls cascading to smaller swirls. Based on all the observational evidence just cited, the argument for supermassive blackholes developing prior to star formation is very plausible. If the supermassive blackholes were created out of the dense Penergy of the early universe, it would mean that the particles were *not* created within the first second following the big bang but created well after the supermassive blackholes began forming. It also means the other tentative deductions made so far appear valid. Additional evidence presented in the next chapter will also support the validity of the first three deductions.

Deduction #4: Considering the observational evidence, it is reasonable to conclude that supermassive blackholes were created out of pure energy from the dense Penergy of the early universe, and their development was well under way when the first particles and stars evolved.

When scientists extrapolate their theories on how the universe will ultimately evolve, they place blackholes as the last things in the universe to exist.[58] This seems appropriate, as it appears they were also the very first things to exist, or at least begin an existence.

Prediction: Someday a bright young physicist will develop the math to properly describe Penergy density. Once that is accomplished, the ability to create a computer simulation of the early universe will confirm that SMBHs can be, and probably were, created by some calculated density of Penergy.

We now have a universe full of emerging SMBHs made of medium-high density Penergy. At the very end of the cascade of swirls the Penergy density thinned to a point where early elementary matter-particles could spin into existence. Let's examine what those first particles may have been like.

Early Particle Creation

The consequence of the creation of emergent blackhole swirls is that the density at the outer edges of those cascading swirls would get thinner and thinner until it was thin enough for objects we call particles to spin into existence. It may seem a stretch to relate the formation of blackholes and particles from the same dynamics, but the idea isn't new.

String theory (*an alternative theory for the physical make-up of particles*) says blackholes and particles may be manifestations of the same basic entity, namely *strings*.[59] Here I am saying that same basic entity is Penergy.

The cascade effect produced several phases of particle creation, and each phase of lighter and lighter particle is commonly referred to as a *generation.* Today we observe three known generations of matter particles, two of which are unstable and quickly decay into lighter versions of themselves. Each generation is substantially lighter in mass than the previous generation. Those observations are consistent with our story of the universe so far.

The first generation of those particles (*the Standard Model labels these Third Generation*) may have spun into existence from light density Penergy, let's say for reference purposes only, at .15 density. As the universe wide Penergy medium thinned further, apparently those particles could not exist in the thinning density and disintegrated into lighter versions of themselves representing the second generation of particles.

That second generation having come about at say .10 density apparently could not continue to exist in that state as the surrounding Penergy density thinned further. They went through another phase transition and disintegrated into the .05 density third generation particles we see today.

It is believed that dark energy (*Penergy*) does not dilute as the universe expands but maintains a constant density.[60] That may well be because the Penergy is thinning so very gradually at this point that it does not appear to be thinning at all. Apparently those .05 particles, the core elements of our Sub-A's, are quite stable in the existing Penergy density medium and will likely stay that way.

The stated values of the densities at which those phase transitions took place are of course only guesses and the actual densities could be quite different. The first and second generation of particles will now only appear when there is a knot of Penergy dense enough to bring them into existence, which happens in our particle accelerators and colliders. Of course, they are in existence only momentarily, not being able to sustain themselves in the now prevailing very thin Penergy density.

Those generations of particles were not actually spun directly out of the Penergy. As we will discuss in the next chapter, the first particles to spin into existence were simple, small, tornado-like balls of Penergy. It is from these early elementary particles that each generation of Sub-A's evolved. There is only modest evidence for the existence of these early elementary particles, but they make sense for several reasons, which we shall examine.

The Cascade Effect would have created many swirls of various sizes that spun into existence, but we only see very large supermassive blackholes and very small, tightly wound particles. What happened to all the swirls sized in between the two?

SMBH/Particle Relationship with Penergy Density

With regard SMBH formation, the larger, medium to high-density swirls would have gained angular momentum and attained sufficient gravitational influence that they would have become SMBHs. As the universe expanded, the Penergy density between swirls would have grown too thin to support the intermediate size swirls having no particle mass and too little gravitational influence to become a blackhole. Being unable to support their bulk in the surrounding, thinning density, they would have eventually disintegrated back into the Penergy medium.

At the other density extreme, the smaller, light density swirls became very-tightly wound entities we are calling early elementary particles. Since they are still with us, they too were obviously capable of sustaining themselves. The small swirls forming the initial generations of particles would have also disintegrated once the surrounding Penergy grew too thin to support their bulk.

There were at least two generations of elementary particles that could not survive in the thinning Penergy density, but possibly there were more. The particles of the first and second generation would be capable of existing longer in the thinning Penergy density when they were bound up with other particles creating a larger composite, but even then, those creations would have limited longevity depending on the density conditions in which they existed.

These transitions would eventually leave the universe with only two stable elements: various sizes of supermassive blackholes and tightly wound, early elementary particles. Both elements still exist today in the remaining very-light density Penergy-medium permeating our universe.

> Deduction #5: We have observed that when particles larger than our Sub-A's are created in particle accelerators and colliders they immediately decay into smaller particles. This implies that those larger particles cannot survive in the very-thin Penergy density level currently pervading our universe and is likely the reason all intermediate sized swirls in the cascade disappeared.

We might speculate that some of the medium to small swirls, perhaps the size of our sun, may have existed long enough and had sufficient gravitational influence to pull some of the early particles with mass inside. This may have fed them with sufficient mass/energy to sustain themselves long enough to eventually become stars, creating the first stars in our universe.

* * *

Starting with only two initial conditions: *the universe started from a speck of pure energy* and that *it began to expand,* our initial observations and deductions have allowed us to build a very interesting foundation to the universe. We first concluded that if the initial speck of pure energy, which we now call Penergy, began expanding then it was likely capable of existing in more densities than its near-infinite state and its very-light particle state. From there we concluded it is unlikely our subatomic particles were created within the first second of the Big Bang as theorized, and that their creation had to await further cosmic expansion and Penergy density thinning.

We then concluded that all the Penergy was not likely consumed in the particle creation stage, leaving a thin density permeating the universe today, which we refer to as dark energy. Since everything we know of in the universe spins, it is likely that at some level of Penergy density large swirls began, setting off a cascade of ever smaller swirls, ultimately ending in only two stable creations: supermassive blackholes and tightly wound early elementary particles.

Given that it appears to be the only way supermassive blackholes could be created within the time frame calculated for their existence, and that the scenario perfectly reflects the *cosmic web* galaxy structure we see in existence today, we concluded that the notion of the Cascade Effect is very plausible.

We have laid the foundation to our early universe — the supermassive blackholes surrounded by a mass of tightly wound, early elementary particles. That is not only a beautiful foundation but one with a great deal of potential. Best of all, it was all constructed from solid observations, without anything exploding or inexplicably popping into existence. Let's move on and see what we can build upon this beautiful foundation.

Chapter Two

The First Building Blocks

So far, the universe has evolved through multiple Penergy states: its near-infinite density state as a pre-Big Bang speck; its medium-high density state as nascent supermassive blackholes; its light-density state as early elementary particles; and its very light-density state as the ever expanding, intergalactic medium. The gradual thinning densities throughout cosmic history may have produced other elements in our universe not yet revealed. Perhaps future discoveries and enigmas will be answered by the concept of a constant thinning Penergy density.

At this point in our story, our expanding universe is comprised of billions of developing supermassive blackholes, billions of disappearing intermediate swirls, and a mass of recently spun early elementary particles. The timing of the swirls becoming blackholes, and the cascade ending in tightly wound particles was apparently just right for the third generation of particles to come into existence without being drawn into the developing blackholes.

Much of the earlier generations may have gone that way, however. We will now look at the universe at this young age and address a problem that is currently plaguing cosmology: Why does the universe today appear so homogenous? Homogenous means something having its components (*like stars and galaxies*) uniformly distributed.

According to Carolyn Devereux, that question is known as the cosmic *horizon problem* (aka *homogeneity problem*) and is one of the most important problems in cosmology.[1] The leading theory to explain the universe's homogeneity has not been made a part of the Standard Model of Cosmology.[2] The theory is rather incredible and deserves our examination. The homogeneity issue also addresses the sequence in which the universe was built, so it is important for our construction project to get it correct. Let's examine homogeneity as expressed in the leading theory and how it would look if the universe were built from the blueprint we've developed so far.

Cosmic Homogeneity

Cosmologists presume the universe is uniform and appears the same from any vantage point within the universe. Over a distance of 300 million light years, from any point in the universe the galaxies and distances between them look much the same, the stars and their contents look much the same, etc. This presumption is so strong it has been named the *Cosmological Principle*.[3] This uniformity, however, is not consistent with the universe envisioned by the cosmological model.

The model does not predict a uniform universe, making the horizon problem a fundamental issue.[4] According to the Standard Model, the star clad galaxies did not start to form until roughly a billion years after the Big Bang.[5] By this time the universe was too big for it to balance out and have its contents uniformly distributed to the degree we see today.

Imagine a large, cold room with a thermostatically controlled wall heater. The temperature drops, the heater goes on and begins to heat the room, creating hot and cold areas as the heat disperses. Given enough time, the temperature in the room would equalize and be fairly uniform throughout. But what if the room was constantly expanding at a rate faster than the equalizing effect could take place. The temperature could never reach a uniformity. This is the scenario scientists have calculated for the early universe. Under the Standard Model's scenario, no matter how far one goes back in time, the universe was never compact enough for a long enough time to achieve homogeneity.[6]

To overcome this and other conceptual problems, Physicist Alan Guth in the 1980's put forth a radical explanation called *inflation.* He theorized that within a few hundredths of the first second of the Big Bang the universe blew up to a comparatively enormous size *very* rapidly. Stephen Hawking described it as if a one-centimeter-wide coin suddenly blew up to ten million times the width of our Milky Way galaxy.[7]

This rapid expansion evidently solves the homogeneity problem but stretches believability. The theory is complicated, involves undetectable fields, the creation of new, undetected particles (*inflatons*), and has other ad hoc characteristics.[8] This is another theory that may work mathematically but is difficult to believe. Tim James in his book, *Astronomical,* points out, 'Inflation is by no means an accepted theory and it generates all sorts of unanswered questions of its own....'[9] Physicist Sean Carroll reminds us that there is no evidence that *inflation* ever happened.[10]

Let's explore an alternative scenario for the universe's homogeneity by looking again at the Cascade Effect of SMBH and particle formation. The process begins with the universe consisting of many proto SMBHs swirling into existence with marginal gravitational influence. As detailed in Chapter One, at the outer edges of the giant swirls, more swirls start their motion, and at their edges more swirls begin, cascading into smaller and smaller swirls, ending in particle generations at the guesstimated .15, .10, and .05 densities.

All this activity took place universe-wide at the outer edges of the cascade of proto-blackholes, which did not yet exhibit sufficient gravitational influence to draw inside much of the newly created particle matter.

The newly minted early elementary particles were free to roam the entire universe and begin combining. Once the bigger swirls gained sufficient angular momentum and a serious gravitational influence, the free particles were drawn closer. As the swirls developed into blackholes, their spin caused the surrounding particles to swirl, flatten to a disc, and form the structure of our galaxies.

All of this suggests the SMBHs and particles completed formation at roughly the same time. Most of the mass of a galaxy is in the content of the surrounding particles and stars. The SMBH generally represents only 1% of the galaxy's total mass.[11] Apparently, they drew inside very little of the early mass of particles but drew near exactly the right amount to support their gravitational field as noted in Chapter One.

This scenario seems consistent with findings of scientists looking at some of the oldest galaxies that date back twelve billion years. As reported in an article dated 1/27/21 in Scientific American entitled *Giant Galaxies from Universe's Childhood Challenge Cosmic Origin Stories,* some of those very old galaxies are too large to have formed in the customary way of being built up from hydrogen gas and star debris. This article is consistent with the one discussed in Chapter One. Both articles say that scientists have no explanation of how those galaxies could have become so big, so soon after the Big Bang.

To explain both the size and ages of the galaxies only requires proper timing for the particles and particle combinations to come into existence just as the pure energy black-holes were gaining gravitational influence. By this time the universe was uniform in its contents and remained homogeneous right up until the time the Standard Model sees atoms created and photons decoupling from matter creating the Cosmic Microwave Background.

The gradual disappearance of the intermediate sized swirls and the gradual growth and development of the larger swirls might account for the miniscule ripples in the Cosmic Microwave Background, discussed further in Chapter Six. This description of SMBH development, particle creation, and the homogeneity of the universe all takes place without the need for the universe to undergo anything as radical as *inflation*.

Because the SMBH creation sequence also easily answers both the homogeneity problem and how the cosmic web was formed, it again validates our first three tentative deductions and allows us to make Deduction#6. It also emphasizes the importance of our understanding the dynamics of Penergy and Penergy density in our universe.

Deduction #6: The creation of pure energy black-holes and the Cascade Effect appear to account for both the formation of the cosmic web and the homogeneity of the universe.

Our construction of the universe has finally brought us to the point of having stable particles in existence. There is still much more construction to accomplish so let's continue by looking at these new particles. So far, the only thing we know about them is that they are tightly wound, little tornadoes of Penergy. Let's see what else our observations can teach us about them.

Early Elementary Particles

Our blueprint for the early universe seems to make sense and is working so far. The natural result of Penergy spinning itself into large swirls that cascade down to tiny swirls is that eventually those tiny swirls would be small enough to spin into what we call early elementary particles. Those particles cannot be today's Sub-A's without some explanation as to how they would have suddenly become so complex.

As discussed earlier, Penergy spinning itself directly into those complex particles just doesn't make sense. We are therefore compelled to conclude that these freshly spun, simple particles are new and different, requiring new names and descriptions.

Inventing new particles is risky for any theory, so proceeding is accomplished with a good deal of trepidation. What follows is not meant to be an exact picture of how the earliest particles evolved but only how they *may* have evolved. Our further construction of the early universe remains as much as possible based on observations and good deductions, but we are diverging significantly from current theory. We will be introducing two levels of particle matter below the subatomic particles we are familiar with today. Don't be thrown off by these new names and descriptions. This picture of the early universe is unique but remains very reasonable and relatively straight forward. It is simply a story about little particles that combine and evolve into more complex particles.

The first particles that spun into existence were truly elementary particles. It means that our electrons and quarks must be composite particles built from these new particles. The argument for composite particles is reasonable and compelling. Consequently, we will proceed with the notion the first, early elementary particles were comprised of nothing more than tightly wound Penergy.

By spinning into existence, these particles would presumably look like tiny tornadoes, so let's call them Torons, or just Tors for short. Everything we say about them will be new and not about any existing Model, so let's call all the related observations and deductions part of a *Tor Model.*

The creation of these new particles ushered in new dimensions to our universe. The creation of distinct bits of particle matter with a finite distance between them implies *space.* Particles flying away from each other at a finite speed implies *time.* These particles were born into an ever-expanding cloud of Penergy; a thinning texture of the interspatial medium we often refer to as *spacetime.*

It is called spacetime because as it turns out the *space* and the *time* are relative to each other, but spacetime is still nothing more than our thin-density Penergy medium. Space and time are relative because time changes whenever space (Penergy) changes. We will examine in depth why space and time are relative in Chapter Five.

We've learned from observations at particle accelerators and colliders that a high energy particle called a photon can decay (*think change*) into particles of matter. This decay process results in a pair of particles, one spinning to the left and one to the right, which fly off in opposite directions. For example, a photon with the requisite energy can suddenly decay into two electron type particles, one being a right-spinning electron and the other a left-spinning positron.[12] From those observations it is reasonable to conclude, as the Standard Model does, that the original creation of particles out of Penergy resulted in the creation of opposite-spinning *particle-pairs.*

Those tornado-like Tors would not be loosely wound like a Kansas tornado. Being created out of Penergy they would be very tightly wound, giving them size, location, mass, and a solid-like composition. Being tightly wound they would not be shaped like an elongated Kansas tornado, but more likely in the shape of a sphere, as we find spinning stars, planets, and blackholes. Created in pairs having opposite spins means they have an orientation of top and bottom, one might refer to as a nose and a tail, respectively. Their spin direction is determined by looking at their tail, down their spin axis as illustrated in Figure 2.1. The point of these observations will be important as we discuss what capabilities and attributes these particles have.

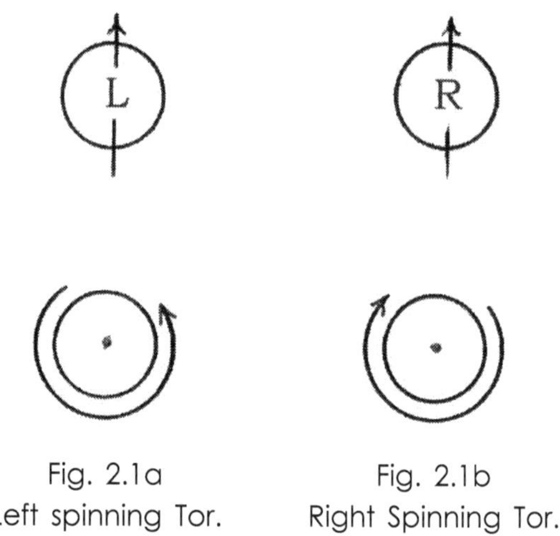

Fig. 2.1a
Left spinning Tor.

Fig. 2.1b
Right Spinning Tor.

Early Particle Attributes

Not being able to see or detect them, we don't know the specific attributes of these early elementary particles, but we might gain a picture of some attributes by making observations about our known Sub-A's. We know Sub-A's spin. This is consistent with our deduction regarding the natural tendency for Penergy to swirl and spin, and our deduction as to how these elementary Tors came into existence. Consequently, we can conclude that Tors spin on an axis.

Spin is very important to the workings of Penergy. It is the primary way for it to change from a cloud into a defined entity with mass, momentum, and other important characteristics we shall discuss in due course. Tors are the only particles spinning on their own axis, which will be important later, as well.

We have observed our Sub-A's to be quite resilient. They can be moderately knocked about like billiard balls without losing their composure or identity. Since Sub-A's and Tors are made from the same state of Penergy, we can conclude that Tors are equally resilient. Perhaps because they are more elementary than the Sub-A's, they may be even more resilient. This is important because the resiliency may be the reason we cannot yet breakdown Sub-A's into deeper levels of constituent particles. The deeper we go, the smaller, more compact, and more resilient those particles are likely to be. We may never see an early elementary particle and may forever know them only theoretically.

We have observed that Sub-A's are known to have *fields*. A field is a disturbance in the Penergy medium creating an area of influence immediately surrounding the particle. Since we have concluded that Tors are tightly-wound, fast-spinning particles like Sub A's, it is reasonable to conclude they too influence the surrounding Penergy medium and therefore have fields. Fields are very important in the construction of our universe, so we will return to them shortly and more extensively in Chapter Five.

As mentioned earlier, when matter particles are created out of Penergy they are created in pairs with opposite spins. Scientists label the pair of particles *matter* and *anti-matter*. The two particles have the same mass and are otherwise alike in every way except for their direction of spin and an attribute called *charge*. They are said to be a *mirror image* of each other.

Scientists have observed that when matter and anti-matter particles meet, they annihilate each other, turning back into Penergy that immediately resolves into two photons (*think particles of energy*). In the early particle creation era that included all generations of particles there was a great deal of particle annihilation and photon production. This would have heated the environment significantly, creating a very hot frenzy of activity.

If matter and anti-matter were created in pairs simultaneously and therefore in equal numbers in the early universe, why is there predominately only matter in the universe today? What happened to the anti-matter?

The Mystery of the Missing Anti-Matter

No one knows for sure where the anti-matter went. It is one of the universe's biggest mysteries. The Big Bang theory speculates that due to an imbalance in the creation/annihilation process in the early universe, out of every billion annihilations there was an imbalance leaving a single matter particle.[13] If this phenomenon happened over and over, it would leave the universe with an abundance of matter particles with few anti-matter particles, which is what we observe today.

That scenario is difficult to believe given the enormous number of matter particles in the universe and the low probability of the same imbalance happening over and over for all known particles. It's also a stretch to imagine the timing of it, since all the matter particle imbalances would have to have occurred within the first second after the big bang. That is a lot of annihilations, creations, and imbalances to have occurred within a single second. The annihilation imbalance theory stretches believability too far. There must be a better explanation.

There are likely other theories for the missing anti-matter, but the least complicated and most straight forward theory is that the anti-matter isn't actually missing. Matter and anti-matter are said to have opposite electric charge values and the universe is said to be charge neutral, meaning equal in charge value.[14] If that is so, for the universe to be charge neutral, the anti-matter must still exist. But if the anti-matter still exists, where is it?

Where is the Anti-Missing Anti-Matter?

The most straight forward answer may not seem feasible at first glance, but it develops into a reasonable premise. If the Sub-A's we know to exist — the quarks, electrons, and neutrinos — are composites of smaller elementary particles, then the missing anti-matter could be a part of the constituent particles that make up the Sub-A's. Anti-matter could be an integral part of the matter we see all around us. Assuming for the moment this is even possible, is there any observable evidence for the hidden presence of anti-matter?

As mentioned earlier, particles of matter and anti-matter are mirror images of each other, differing only in the direction of their spin and charge.[15] Charge can be either negative or positive; attributes ascribed to matter and anti-matter respectively. To find anti-matter therefore we only need to hunt down particles carrying a charge opposite of that of matter. But we know of those particles already. Electrons have negative charge and protons have positive charge. Up quarks have positive charge and down quarks have negative charge.

If matter and anti-matter are known to carry opposite charges, and we find both types of charge existing in the particles around us, it suggests both types of matter exist respectively within those particles. If down quarks are negatively charged and referred to as matter, it makes sense that up quarks that are positively charged contain at least some anti-matter. These are bold ideas with some obstacles to overcome but given no better explanation for solving the anti-matter mystery they are worth examining.

If our Sub-A's possess both types of matter it would mean our Sub-A's are definitely composite particles and that composite Sub-A's would solve the missing anti-matter mystery. There are good arguments for the existence of composite particles, and they have many other benefits as we shall see. Let's proceed with both notions: that the first particles to spin into existence were true elementary particles from which Sub-A's evolved, and their constituent particles might include anti-matter particles.

The idea that our Sub-A's may not be true elementary particles is not new. Several physicists have suggested that deeper levels of elementary particles are possible. Physicist Jon Butterworth tells us both categories of matter particles, quarks and leptons, may contain smaller constituents.[16] Physicist Kenneth Ford tells us that even the lowly neutrino particle may not be a pure particle, but a mixture of two other particles, each with a definite mass.[17] Chemical Physicist Michael Munowitz tells us that quarks may not be the indivisible, end-of-the-line building blocks, but they are probably only a step or two away.[18]

With these notions in mind, let's continue with our construction project. First, we must consider our biggest obstacle: how does one construct composite particles by mixing matter and anti-matter together without them annihilating?

Spin Signature

Annihilation occurs between colliding particles when the two are mirror images of each other as are all matter/anti-matter pairs. It is unknown why matter and anti-matter annihilate; perhaps the opposite spins cause the two particles to unwind each other when they meet.

An electron and its mirror image, the anti-electron, will annihilate, but an electron and an anti-muon (*a second-generation version of an anti-electron*) will not annihilate.[19] Either will an electron and anti-proton.[20] In each case the two particles have opposite spin and charge, but apparently the mass difference between the respective particles precludes annihilation.

In another example, a subatomic particle called a meson is made up of a quark and an anti-quark. The two quarks can co-exist together without annihilation because of a characteristic called *color* that differs between the two. Apparently, the slight difference between those examples is enough to preclude the usual matter/anti-matter annihilation.

It seems that in order to annihilate, the two particles must be perfect mirror-images of each other, in perfect symmetry, and a deviation in that perfection will preclude annihilation. But how does an electron know whether it is a mirror image of an anti-muon? There must be something inherent in each particle that allows it to identify its mirror-image.

Particles have few primary traits — mass, spin, and charge. For mirror-image particles to annihilate, we know based on the examples above that the mass must be the same between the particles, and we know the charge must be opposite. That leaves only spin as the inherent, identifying trait. It would mean that each particle type has its own special spin, or *spin signature*, and only particles with the exact mirror-image spin signature can annihilate.

The difference in spin signature preventing annihilation might be something subtle such as a difference in the wavelength (*think energy level*) between the particle and anti-particle. Michio Kaku reports that in a very cold trap to store anti-protons in a magnetic field, the wavelength of the cold anti-protons would be much longer than the wavelength of the protons of the atoms in the container walls, so the anti-protons would reflect off the walls without annihilation.[21] Here are touching, mirror-image particles, but their respective energy wavelengths apparently constitute a significant deviation creating slightly different spin signatures, thus no annihilation.

Deduction #7: Given that particles with opposite spin and charge don't annihilate when their masses or other traits are different, we can conclude there is an inherent trait that allows particles to identify their perfect mirror-image counterparts. We have identified that trait as *spin signature* and concluded that only mirror-image particles with the same spin signature annihilate.

Prediction: Experiments involving different configurations or states of anti-matter will confirm the validity of spin-signature, or a similar theory, paving the way for a complete, composite particle theory.

We are set to continue construction of our early universe with matter, anti-matter, and spin signature elements in our toolbox. First, we need to fortify our blueprint with a clearer image of what is happening in the big picture of this project. What is shaping our construction? In other words, how is this cosmic evolution progressing?

Evolution Progresses in Levels

There are many theories about evolution, each having some validity depending on their perspective. One such perspective is to think beyond planet earth; to step back and observe how matter has evolved cosmically over the past several billion years.

Going back to the evolutionary level we recognize as **Sub-A's**, scientists have theorized that in time the existing positively charged protons and neutrally charged neutrons *combined* to form nuclei. The nuclei eventually *combined* with various configurations of negatively charged electrons creating the next evolutionary level of matter — a variety of **Atoms**.

Once the atomic level stabilized, the atoms went through their own evolutionary stages, *combining* to form hydrogen gas molecules that in turn *combined* to form stars, and later stars and planets. On at least one planet (*earth*) molecules of atoms *combined* to form large macromolecules, creating proteins, carbohydrates, nucleic acids, and lipids, that eventually *combined* to form the next evolutionary level of matter — the **Cell**.

Cells in turn went through their own evolutionary stages developing sensory apparatus, locomotion features, organs, a circulatory system, a nervous system, and the ability to *combine*. These advancements lead to the creation of the next evolutionary level of matter — the **Organism**.

Based on observations of just these four evolutionary periods we might conclude that evolution progresses in stages to *Levels*, and that each Level is the result of the previous Level growing in complexity and *combining*. We could call this process the Combination and Growth Process. Since it has existed for billions of years and seems very precise in both its persistence and direction, if we can find evidence it has existed since the birth of the universe, we could more precisely call it the *Combination and Growth Imperative*, or the CAGI for short.

The constant building of complexity has long been observed by others. Physicist Paul Davies tells us, 'The fact that nature has *creative power*, and is able to produce a progressively richer variety of complex forms and structures, challenges the very foundation of contemporary science.'[22]

His book, *The Cosmic Blueprint,* was published in 1989, but the notion is still relevant today.

To examine whether this CAGI theory has merit we must determine whether it is feasible for the early elementary particles to have somehow combined to get that combination and growth process started. But Tors are simple, spinning tornados of Penergy; how can they possibly combine into something meaningful?

The Emergent Qualities of Tors

We will likely ask the same 'how could this possibly happen?' question at the start of each evolutionary Level because the dominant structures at the start of each Level always seem so simple, they show no evidence or promise of combining potential. What we later observe however, is that each Level evolves new and exciting *emergent qualities* that will assist it in changing, growing, and combining.

Emergent qualities come about according to Physicist Stephon Alexander when elementary constituents interact to create novel properties that are not possessed by the constituents themselves.[23] For example, looking at two hydrogen atoms and one oxygen atom, one could not anticipate their combining to produce a substance such as water. Nor could one anticipate the creation of all the other gases, liquids, and metals that can be produced from other combinations of simple atoms. Emergent qualities are at the heart of the CAGI and the evolutionary process.

In addition to the basic qualities already mentioned, Tors have emergent qualities that are very important not only for their future but for the future of every level of matter to forever evolve. These qualities involve the *forces* that hold matter together.

If particles are to combine, they must interact. Any interaction between particles according to current theory is a function of *forces*. Scientists have theorized four forces working in the universe. The Strong and Weak forces only involve Sub-A's confined to the nucleus and will be discussed more fully in Chapter Six. The Gravitational force is the attractive force between masses that holds us to the surface of the earth, but it is far too weak at the particle level to be a combining force.

The Electromagnetic force is a combination of the Electric force that pushes electricity and the Magnetic force that draws magnets together. The electric force involves two different charges, positive and negative, that may be inherent in two different particles. The magnetic force, however, might emanate from a single type of particle. Let's examine the magnetic force to see if it could be a candidate for the force holding Tors together.

The Magnetic Force

Scientists have observed that when magnetic attraction comes into play the particles involved are lined up in the same direction, meaning the axes of their spins are aligned.[24] A natural magnet called lodestone (*magnetite*) is *magnetized* by virtue of the electrons in its atoms having their spin axes aligned. When the lodestone is placed near a metal object it causes the spin axes of the metal's electrons to align in the same, parallel position, *magnetizing* the metal.

The magnetic body with its electrons so aligned is said to have a head and a tail, or a north pole and a south pole. When the north pole of one magnet is placed in line with the south pole of another magnetized object, they are attracted to each other. But when they are aligned otherwise, especially north to north or south to south, they repulse each other, as you know if you've ever played with magnets.

But what is causing this unseen force aligning the spin axes? Where is the force coming from? The force was once believed to be an undefined inherent aspect of certain pieces of matter, namely metals. The Standard Model has its own explanation for forces involving the exchange of virtual particles, and the magnetic force simply being a part of the electromagnetic force, which will be discussed in Chapter Five.

The Tor Model sees force differently and ascribes the magnetic force to the *field* of the Tor particle. The Tor particle spin causes the surrounding Penergy to spin with it and spread out creating the field. When the spinning field of one Tor comes into a nose to tail alignment with the spinning field of another Tor particle, they are attracted to each other. If the spin alignment is otherwise, the particles repulse each other. This is the basis for the magnetic force.

The force is evident in our observations of a *magnetic field* represented by iron filings on a paper placed on a bar magnet, as shown in Figure 2.2. The magnetic field lines loop through the north and south poles, entering through the south pole and exiting through the north pole. When two bars are placed in line, those field lines continue through both bars creating a natural attraction between the north and south poles. We also observe that the strength of the field is greatest near the poles and falls off as one moves away from those poles.

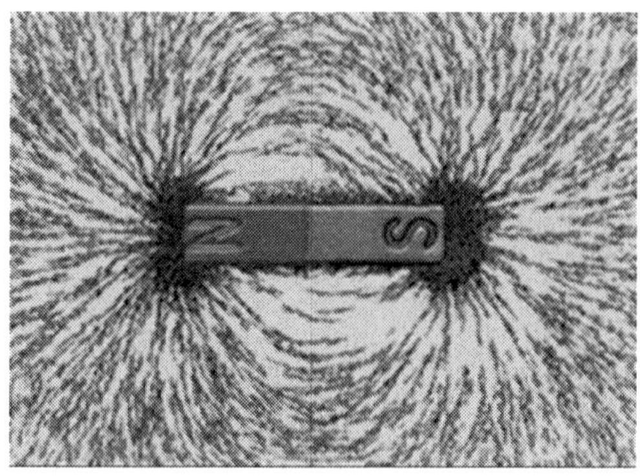

Figure 2.2 Iron filings on a sheet of paper covering a bar magnet showing its magnetic field.

This picture of iron filings is a good example of particles producing a field around themselves, and the cumulative field influencing other particles. In this case the aligned atoms within the magnet are producing a magnetic field that is in turn influencing the atoms in the iron filings, causing them to move into a fixed orientation consistent with the magnet's field.

In summary, the fields of two Tors will cause the Tors to be attracted to each other if their fields are aligned nose to tail, and repulse each other if meeting in any other arrangement, especially if nose to nose or tail to tail.

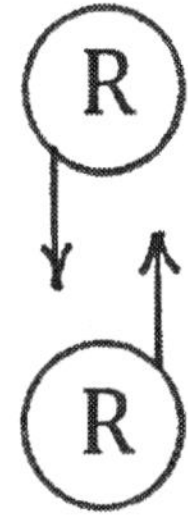

Fig. 2.3a
Tor magnetic *attraction* occurs when like-spinning Tors are aligned nose to tail.

Fig. 2.3b
Tor magnetic *repulsion* occurs when the Tors are otherwise aligned, here shown aligned tail to tail.

The Creation of Tor Chains

The magnetic force is a very reasonable answer to what can bring Tor particles together. If so, when aligned nose to tail they are attracted with sufficient strength to hold themselves in that attached relationship. Such in-line attachments would create a *chain* of like-spinning Tors, as illustrated in Figure 2.4.

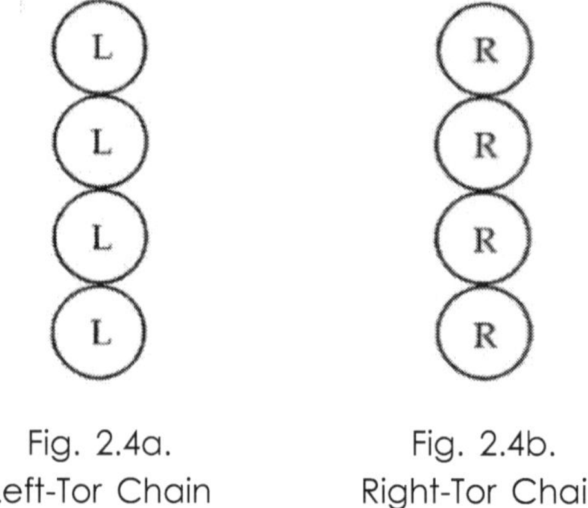

Fig. 2.4a.
Left-Tor Chain

Fig. 2.4b.
Right-Tor Chain

Tor chains can be left-spinning or right-spinning and can be any number of Tor particles long. Longer chains would make it less likely the chain would be annihilated by bumping into an opposite spinning chain of the same length and spin signature. However, the longer the chain the more likely it would be broken apart by a collision with a more energetic Tor or Tor chain.

> Deduction #8: The source of the magnetic force may be inherent in the Tor-particle field, causing an attractive force when Tors are aligned nose to tail, which allows for the creation of Tor-Chains.

A Tor chain could be hundreds, even thousands of Tors long. In the early universe these long chain creations probably happened, but the utility of such long chains seems doubtful. On the other hand, proteins are long chains of nucleic acids encompassing thousands of atoms that have proven to be very useful if not essential to evolutionary growth. Tor-Chains may turn out to be equally essential to evolutionary growth and we should be open to how many Tors make up a viable/useful chain.

* * *

We have built the first layer of our universe upon the foundation of supermassive blackholes and particles, which we will call the Tor-Particle layer. They represent the first true building blocks to our physical universe.

We took notice of how easily the homogeneity of the current universe was created, without the need for anything as drastic as *inflation.* Bringing early elementary particles into existence is risky for any model because we have no solid evidence for their existence and must rely on circumstantial evidence and reason. But their origin by spinning into existence and evolving into today's subatomic particles makes more sense than having them simply pop into existence without rhyme or reason.

Composite particles answer the mystery of the missing anti-matter and provide a better origin for the magnetic force. We concluded that the magnetic force may well be responsible for bringing the first early elementary particles together to form Tor-Chains.

Stepping back momentarily to view the bigger picture of our project, we saw that we may be building ever more complex forms with ever more complex emergent qualities. This gives rise to the possibility for cosmic evolution being driven by a complex combining force we might call the *Combination and Growth Imperative,* or the CAGI for short. We shall pay attention to that combining process in the chapter ahead to see whether the notion of the CAGI holds together.

We have taken the first step in combining particles to create composite particles. Our wild ride through the particle creation era is exciting and continues in the next chapter with the construction of the next building blocks of our universe.

Chapter Three

Level Two Building Blocks

The evolution of our early universe has been quite a journey already. Our construction project has undertaken many steps and we are still working on the building blocks of the first level — the Tor Level. Our universe is now filled with developing supermassive blackholes, cascade remnants of faltering swirls, and Tors — the early elementary particles capable of surviving in the thinning Penergy medium.

The Tor fields, endowed with what we now call the magnetic force, would have allowed Tors to combine nose to tail and create various lengths of Tor-Chains. If cosmic evolution progresses through the Combination and Growth Imperative (*the CAGI*), the individual Tors and Tor-Chains will be compelled to combine into larger, more complex, stable particles. But they are only simple particles struggling to avoid annihilation in a hot frenzied environment; how are *they* going to combine?

Particle Assembly from Tor-Chains

Mirror-image matter particles have a natural attraction for each other, e.g., electrons and anti-electrons are observed to attract each other.[1] It is therefore reasonable to conclude that mirror-image elementary Tors have a natural attraction for each other. We can also conclude that mirror-image Tor-Chains would likewise have a natural attraction for each other. Exact length mirror-image Tor-Chains would annihilate, but if they had different lengths, they would have different spin signatures and therefore would attract each other without annihilation.

That was a lot to take in, so allow me to repeat it for clarity. Opposite-spinning Tor-Chains have a natural attraction for each other. Chains of different lengths have different spin signatures and therefore do not annihilate. The attraction allows opposite-spinning Tor-Chains of different lengths to successfully combine, as illustrated in Figure 3.1.

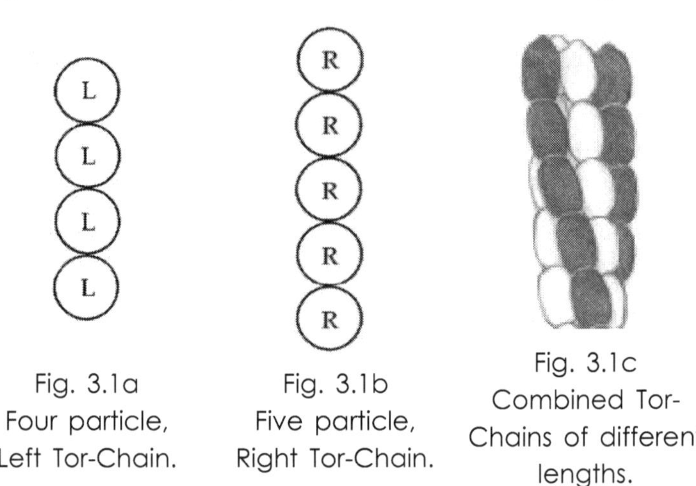

Fig. 3.1a Four particle, Left Tor-Chain.

Fig. 3.1b Five particle, Right Tor-Chain.

Fig. 3.1c Combined Tor-Chains of different lengths.

Figure 3.1. Left and Right Tor-Chains of different lengths have a safe attraction for each other and combine without annihilation.

Tor-Chain attraction could create combinations of various chain lengths with any number of chains involved. It may have been such a combination of various chain lengths that created the first modern particle, the *photon*, whose configuration is much like that shown in 3.1c. The photon configuration will be discussed in more detail in the Chapter Four.

Particle Assembly from Tors and Tor-Chains

Again, individual mirror-image Tors will annihilate, but when bound up tightly in a group such as a chain, the individual Tors are no longer subject to annihilation. Only a mirror-image chain of the same length and spin signature can annihilate the chain. The environment of many different chain lengths slows the annihilation rate. This allows the chains to survive long enough to combine and become a part of a larger composite.

Having Tor-Chains and single Tors available to combine, the next stable combination might be a long Tor-Chain *encircled* by any number of single Tors. For simplicity that notion will be illustrated in Figure 3.2 as a four-Tor-Chain encircled by a single, opposite spinning Tor. All kinds of different combinations are of course possible. We will call this individual-Tor and Tor-Chain combination a **Tryk**. The mirror-image form of the Tryk configuration would also evolve and be stable, and as usual if the two should meet they would annihilate each other.

In a Tryk, the single, opposite-spinning Tor has the same circling dynamics as that of an electron encircling a nucleus to create an atom. That dynamic is the electric force wherein matter-particles with opposite *charge* are attracted to each other. The electric force will be discussed further in Chapter Four.

Given that Tors can combine into chains, and opposite spinning Tors and Tor-Chains can attract each other, it is possible to have all sorts of combinations evolve if they are not mirror images of each another. We could have a twenty-Tor-Chain encircled by a single Tor, or a one-hundred length

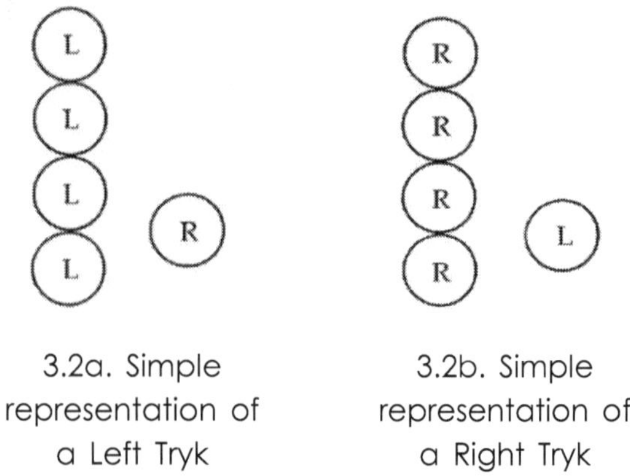

3.2a. Simple representation of a Left Tryk

3.2b. Simple representation of a Right Tryk

Figure 3.2. Representations of a Left and Right Tryk made from a Tor-Chain encircled by a single, opposite spinning Tor.

Tor-Chain encircled by twenty Tors. All kinds of combinations are possible.

The creation and destruction of all those possible combinations would be on-going and constitute the evolutionary phase of the next building blocks. Many combinations would come into existence, be broken apart, recombine into different configurations, and be broken apart again. It's the same product strengthening process every evolutionary Level goes through before a stable, dominate form eventually evolves from the frenetic environment. This evolutionary process establishing which combinations survive and go on and which do not is called *natural selection.*

Survival by Natural Selection

The actual combination of Tors and Tor-Chains in the configuration of Tryks is unknown. They are probably complex and those illustrated in Figure 3.2 are only put forth to show what might be possible. With all the creation, annihilation, and combining of elementary particles taking place, the thin Penergy medium would have heated up, becoming a hot, high-energy *frenzy* of activity. Only the strongest particle combinations would survive the frenzy. In this early stage of cosmic evolution, survival by natural selection had begun.

Biological natural selection is the process of organisms undergoing physical change due to DNA mutations. Some mutations make the organism better adapted to its environment, which allows it to survive and reproduce, perpetuating those survival traits. In the case of elementary particles, the strongest and most suitable composite combinations would tend to survive the frenzied environment and go on to become stable and available to combine again.

Collisions breaking up weaker connections that reconfigure themselves is the mutation equivalent for particles. Which Tors, Chains, and Tryk combinations would survive annihilation or be broken apart by collision is impossible to say, but natural selection in this hot, frenzied environment would definitely determine their number and composition.

We have now created two Levels of building blocks toward building our early universe and we haven't even reached the Level of Sub-A's, though they are next. In the Standard Model the Sub-A's are the elementary particles. In the Tor Model the Sub-A's are *composite* particles comprised of Tors and Tryks. Let's examine the idea of composite particles and determine whether they have any virtues supporting the argument for their existence.

Virtues of Composite Elementary Particles

Penergy spinning itself into various generations of Tors seems to be the most likely way for particles to have come into existence. Penergy might spin itself into a pair of complex Sub-A's, but simple elementary particles that then combine to create Sub-A's makes more sense. Many physicists agree that our Sub-A's may have a substructure of more elementary building blocks. Physicist Harald Fritzsch tells us that though there is no experimental evidence for it yet, it is possible that the electron possesses an inner structure.[2]

Jon Butterworth points out that what we think of as fundamental particles might simply be the smallest things we can measure with the tools available, and there may be deeper layers.[3] Stephen Hawking told us that at very high energies we might expect to find several new layers of structure more basic than the quarks and electrons that we now regard as elementary particles.[4] Jorge Cham and Daniel Whiteson tell us there is no proof that the particles we see today, electrons and quarks, are the most basic building blocks in the universe, but are [*merely*] the smallest bits of matter we have seen so far.[5] The following points support the notion our Sub-A's are composites of more elemental particles.

- As we have seen, composite particles may very well account for the missing anti-matter. The anti-matter is not only a part of composite particles, but an essential part. Anti-matter's natural attraction to matter makes it possible for Tryks to exist. As we progress in constructing future building blocks to our universe, the anti-matter particles will again prove to be an essential element in that construction.

- Composite particles could be the underlying source of *magnetism* and *charge.* As discussed in Chapter Two, Tor-Chains with all their spin axes aligned create magnetic fields that encourages other particles to align their spin axes accordingly. As we will discuss in the next chapter, the combination of elements in composite particles will explain charge and why some Sub-A's have differing amounts of charge.

- Composites imply the existence of a constituent core particle such as the Tor. The validity for the existence of a particle like the Tor is implied in much of the theories and speculations discussed so far. We will soon speculate on the spinning of the Tor being the source of mass. In Chapter Five we will speculate further on how the Tor could also be in addition to being the source of the magnetic and electric fields, the source of the *gravitational field.* It seems like quite a bit to expect from a single particle, but it makes more sense for these attributes to arise out of the most elementary particle than for those forces, according to the Standard Model, to have inexplicably popped into existence.
- The Standard Model recognizes particles to be very consistent in their attributes, such as charge and mass, but offers no credible explanation as to why this is so. On the other hand, the idea of evolving, composite particles supports sound reasoning for consistent particle attributes, which will be discussed in the next section.
- The creation of composite particles infuses natural selection into the evolutionary process, which seems a much more likely influence on particle composition than the *it just happened* approach taken by the Standard Model.

Scientists tell us that all charges are rational multiples of a basic unit of charge.[6] The fact that there are three Sub-A's (*electron, up quark, and down quark*) with the same type of electric charge but with different values makes it probable they share a common, underlying structure. Jorge Cham and Daniel Whiteson tell us that because the charge of the electron perfectly matches the charge of the proton but is opposite, it is another sign there are deeper components underlying those particles.[7]

All these facets weigh heavily toward today's Sub-A's being composites of more elementary particles, named in the Tor Model as Tors and Tryks. Based on these arguments and what we have discussed earlier we can make our next deduction.

> Deduction #9: Given that Penergy spun itself into a cascade of swirls ending in early elementary particles; and given those particles can combine due to magnetic and electric forces; and given the many supporting points for our Sub-A's being composite particles; we can conclude that today's Sub-A's are composite particles made up of elementary particles like Tors and Tryks.

Before moving on with constructing the next level of building blocks, let's examine more of the attributes of the particles created so far.

Elementary Particle Attributes

Particle Levels and Size

In the Tor Model, evolution progresses by Levels through the combination and growth of matter (*the CAGI*). Tors represent the most elementary level, so they are referred to as Level 1 particles. Tryks (*or something like them*) are built from Tors and represent Level 2 particles. Tryks will be the building blocks for the Level 3 particles, the Sub-A's we are familiar with today. There might be an intervening level of particles, but for simplicity let's continue with what we have and build our Sub-A's from the Tryks illustrated above. Other scenarios are of course possible.

Tryks are very tiny, tightly wound particles. We may never see them or be able to prove their existence. Composite particles hold themselves together by a force through their fields, the strength of which is called their *binding energy*. It takes at least the amount of the binding energy to pull apart the constituent elements of a particle. Tors spinning on their own axes create a very strong field. Due to the strength of the Tor fields making up a Tor-Chain, the binding energy of Tor-Chains and Tryk-Chains is very high. We may never build a particle accelerator with sufficient energy to overcome that binding energy to break matter down to that level.

Tryks can combine to form particle chains of any length. Still quite small, these chains would look like minute bits of wiggling string. The wiggling, spinning, vibrating movements of each particle affects the Penergy around it, creating a field. The field in turn affects the behavior of other particles passing close by. In this way particles affect and possibly even communicate with each other. These are all important attributes of early elementary particles, characteristics that will prove essential to future Levels of evolution.

Particle Consistency

As far as scientists can tell, all electrons possess the exact same mass and electric charge value. All protons possess the exact same mass and electric charge value. Those particles created anew out of knots of Penergy inside colliders have the exact same mass and electric charge value as their counterparts that were created billions of years ago. What can account for such consistency?

Third generation Tors were created within a *band* of Penergy having a guesstimated .05 density, producing Tors with slightly varied masses. Only Tors of the exact same mass can combine (*or annihilate*), so once a certain size Tor was eventually *naturally selected* to be dominate it would create consistent Tor-Chains and Tryks.

The other Tors and Tor combinations would eventually be annihilated, torn apart, or become part of the fragments of particles making up *dark matter,* which will be explained in Chapter Five. One naturally selected Tor size also assured that subsequent particle types created from Tors and Tryks would have the exact same mass, charge, and other overall characteristics. In the end, particle consistency is a product of natural selection.

Particle Mass

Penergy does not have *mass* as we define it until it is spun into a blackhole or particle, which then gives that Penergy content, definition, and location. Particle mass is much different than blackhole mass, which will be discussed in Chapter Five. Mass is one of the core characteristics of a particle and is a measurement of the amount of Penergy comprising the particle.

Because of Einstein's famous equation $E=MC^2$ and other related equations, mass is often expressed in terms of energy and vice versa. Physicists use a single term called the *electron volt* (eV) as a measurement for both mass and energy. An electron volt is the amount of energy gained by an electron moving across a one-volt, electric field.

To be consistent with current scientific terminology of which the reader may already be acquainted, the eV will be used to express mass. MeV stands for a *million electron volts,* which is the typical value used for most particle masses. If the reader is challenged by the term electron volt don't worry, it's only a measurement term. Think of it like other measurement terms such as an *inch* or a *pound* that are used simply to state a value. Here the term is used for comparative purposes and its exact value is unimportant.

The Standard Model posits the existence of a field you may not have heard of called the *Higgs Field.* It is named after Physicist Peter Higgs who was instrumental in creating the theory for its existence. It is a field that supposedly exists throughout the universe and gives mass to particles. In other words, it is the particle's interaction with this undetected field that theoretically causes particles to have mass.

The field is theoretically created by the Higgs boson particle. This boson is a very large (*125,000 MeV*) particle that has not actually been seen but is believed to have been brought into existence at the Large Hadron Collider (LHC). Inside the LHC it survives for only 10^{-22} of a second.[8] That is less than a millionth of a millionth of a billionth of a second. The creation of the Higgs boson is a rare occurrence happening within the LHC only once in every ten billion particle collisions.[9]

Again, the Higgs boson has not actually been seen. Its existence is implied by reading the decay debris detected in the collider.[10] Reading debris created in particle colliders is notably imprecise. The process has been compared to throwing a piano out the window and then trying to determine all the piano's properties by analyzing the sound of the crash.[11]

The Higgs boson was initially predicted to be much larger. For the calculations to come within the vicinity of the mass observed, the mass number had to be *fine-tuned* (*think mathematically manipulated*).[12] Despite these problems, the Higgs boson was reported discovered in 2012 at the LHC.

The Higgs boson maintains the Higgs field, yet the boson decays quickly, so it does not naturally exist in our universe. Particles theoretically gain mass by interacting with the otherwise undetectable and ubiquitous Higgs field in a way that slows the movement of the particle, which causes it to act like our everyday understanding of mass. According to Physicist Lawrence Krauss in his recent book *The Greatest Story Ever Told ... so far,* if the Higgs field can make a particle more sluggish, the particle *acts* like it has a commensurate amount of mass.[13]

The problem with sluggishness equaling mass is that it is inconsistent with Einstein's E=MC². According to Einstein the mass of a body is a measurement of its energy content. Apparently in the Higgs field the particle does not actually contain the requisite energy represented by its mass, it only *acts* like it does. There is a mathematical structure to support the theory, but apparently there is no actual evidence the Higgs field exists, except by the presumption that if a particle exists then a corresponding field exists.[14]

Harald Fritzsch tells us that physicists invented the hypothetical Higgs field simply out of the need to give particles mass.[15] Astrophysicist David Lindley tells us the Higgs field explanation for mass is ingenious, but it is a trick, a gadget, a kludge, as computer programmers would say of a piece of code that is tacked on to a piece of software to perform some necessary but overlooked task.[16]

The Higgs boson may be everything the physicists say it is but producing a knot of medium-light-density Penergy by slamming protons together will always produce a big, unstable particle because that is how Penergy works. Predicting a big particle will be produced and decay into smaller particles that will decay into yet smaller particles, is not convincing evidence for a particle producing an undetectable field that somehow gives particles mass. Slamming high energy particles together to produce a knot of Penergy and a cascade of lighter particles only proves the presence of our thin-density Penergy and the instability of heavy particles created within it.

The Higgs theory may satisfy a theoretical need, but it is difficult to believe and not foolproof, so particle mass could just as well have more to do with the density phase of the Penergy when the particle was created, as observed in Chapter One.

Knowing exactly how particles got their mass may not be essential to our building the early universe, but given the problems inherent in the Higgs notion, let's speculate on an alternative idea. Let's start with two associated observations regarding the nexus between mass, spin, and spacetime.

The first is that particle interactions have shown particles to have momentum, which means they have mass, which is known through our understanding of gravity to distort spacetime. Secondly, we are told that fast spinning bodies such as blackholes also distort the surrounding spacetime.[17] If mass can distort spacetime, and spin can distort spacetime, then perhaps what is actually distorting spacetime isn't mass but something commensurate with mass, such as the *spin* of the particles making up that mass.

An Alternative Source of Mass

Tors are the only particles spinning on their own axis and being pure Penergy presumably do so at a *very* high rate. Things that spin on an axis at a high rate create a gyroscopic effect. That means that the spinning body is inclined to stay in the same position relative to its spin axis, resisting any movement or tilt to that spin axis. You may recall this phenomenon if you have ever played with a toy gyroscope.

Physicist Frank Wilczek describes elementary particles as ideal, frictionless gyroscopes that never run down. He also points out that the faster a gyro rotates, the more effectively it will resist attempts to change its orientation.[18]

Perhaps a body's resistance to a change in movement or direction, which is how we measure a body's *inertial* mass, is simply the resistance inherent in many Tors acting like mini gyroscopes. The more spinning Tors making up the body, even in all sorts of orientations, means more resistance to movement and hence more mass. Only in a situation where all the spinning Tors are perfectly aligned and balanced would one find a *massless* body such as in the configuration of a photon, which will be discussed in Chapter Four.

The above speculation may not be correct, but it makes a point. This speculation has the idea of mass emanating out of a natural attribute of the most elementary particle, the Tor. By building the universe on the premise our Sub-A's simply popped into existence directly out of the speck of energy, the Standard Model has gone down a mathematical path that has passed up opportunities to understand the universe in a more elementary way.

The source of mass might well have something to do with the most elementary particle, but if we don't recognize the most elementary particle, that fact will never be explored. The point is, if the math is taking us down a path of questions without reasonable answers, it is time to stop and look back up the path to see where we may have taken a wrong turn. As observed in the introduction, math can be a wonderful tool, but it is plagued with the potential for wrong turns.

How Particles Combine

Tryks are the stable, second level of matter. To reach the third level, Tryks will have to combine to create a more complex form. In the Tor Model, particles combine in two ways. Firstly, due to a natural attraction inherent in their fields such as the way like-spinning Tors combine nose to tail to create a chain, or the way the natural attraction between electrons and protons creates atoms. Secondly, two or more particles can also combine by sharing parts of themselves such as protons and neutrons sharing Pions and

Gluons to create a nucleus, or atoms sharing one or more electrons to create a molecule.

Either of these methods will hold two or more particles together allowing for the creation of larger particles. In our examples here, Tryks combine using natural attraction, but other scenarios are possible. According to the Standard Model, protons are made up of three quarks that are held together by a strong force that involves both sharing and natural attraction. Such a force may represent an entirely different way of combining, but its existence is only theoretical and not completely understood.

The third level of matter will emerge from the frenzy of the environment to be quite complex compared to the first two Levels. For the third Level to evolve, many kinds of Tryk combinations will be made and torn apart, with their fragments recombining to form yet other combinations. Our Penergy medium will have become a sea of Tors, Tryks, and their combinations, much like the earth's oceans were a sea of macromolecules prior to cells evolving.

Figure 3.2 shows two Tryks that can be used to build more complex particles, which will be covered in Chapter Four. Again, the actual combinations of elementary particles whatever they look like are probably much more complex, and these simple combinations are put forth only to illustrate what such combinations *might* look like.

All this particle creation, annihilation, and smashing combinations apart creates a Penergy environment that is truly a *frenzy.* It's a wonder any combination of Tryks make it, but given enough time, pushed by the CAGI and tested by natural selection, viable combinations will prove stable enough to survive. We now turn to examine how Tryks might combine to create Sub-A's.

* * *

Recognizing we can only guess at what the early elementary particles might be like, we journeyed forward with the understanding we are building particles only from representatives of what they might be. We continued our construction project by combining Tor-Chains and then by combining Tor-Chains and individual Tors to create a new Level of matter called Tryks.

We examined the importance of Natural Selection and the part it plays in assuring each phase and Level of evolution is capable of both surviving within its environment and possessing the capacity to combine. We then examined the virtues of composite particles and concluded that our subatomic particles are most likely composites of early elementary particles. We looked at what the attributes of the early elementary particles might be like and speculated on an alternative theory as to how particles got their mass.

Chapter Four

Level Three Building Blocks

Our universe is well under way and is beginning to look a bit more like the universe we are familiar with today. The supermassive blackholes are taking hold, the intermediate swirls have all but disappeared leaving the skeleton of the cosmic web, the elementary particles are combining, and our Sub-A's are about to evolve from Tryks.

Tryks brought important emergent qualities to the universe. The fields surrounding the Tryks would reflect the fact they are comprised of both Tor-Chains having a magnetic field and a combination of matter and antimatter creating an electric field. This combination of essential emergent qualities together creates the electromagnetic field, which gives particles the ability to absorb and emit photons.

This gives the Tryks the capacity for differing energy levels to promote various interactions and combinations. These emergent qualities not only assisted the Tryks in combining, but they would also prove essential to all future levels of matter as well.

The cosmic environment remains a hot frenzy of activity, so these new particle combinations must be resilient to survive. The dominant form of the Tryk may not have completely evolved yet, so many trial combinations of Sub-A's would have been created, testing the combining strength of the Tryk. This slow, testing transition occurs between all Levels; it is simply a part of the CAGI's evolutionary process. We will proceed as if the Tryks appearing in Figure 3.2 were the dominate form that successfully evolved, but of course the actual configuration was probably much different.

Subatomic Particle Construction

Scientists use large particle colliders such as the twenty-seven-kilometer, circular, Large Hadron Collider (LHC) located underground near Geneva, Switzerland to smash Sub-A's together at high speed. The collision creates a knot of Penergy big enough to bring into momentary existence larger, unstable particles that immediately decay into smaller particles. Scientists have developed the detection equipment to record the behavior of these smaller particles, which gives the scientists the basis for theorizing the particle's existence, characteristics, and relationship to other particles.

Some particles are so fleeting they have not yet been seen but are theorized based on the behavior and characteristics of other particles. Scientists have identified four, *stable* particles: the quark, neutrino, electron, and photon. They have also either seen or theorized a host of unstable particles and anti-particles.

All matter as far as we know is made from quarks, neutrinos, and electrons. Photons are considered energy particles that provide matter particles with the requisite energy to move, interact, and combine. Due to shared features, electrons and neutrinos have been placed in a subcategory called leptons. There is also a particle category of virtual particles (*think briefly popping in and out of existence*) called Bosons that according to the Standard Model transmit force. The Tor Model views the source of force differently. The source of the electric force is discussed below, and we will discuss force again in Chapter Six.

The Standard Model views quarks and electrons as elementary particles, but again some physicists believe they may not be the most fundamental particles. Harald Fritzsch informs us that quarks and leptons might consist of smaller building blocks.[1] Physicists John Barrow and Joseph Silk tell us that physicists would not be surprised to discover ultimately that quarks and leptons have internal constituents.[2]

Let's see how these quarks, electrons, and protons we are familiar with today might be created from Tryks. We begin by examining the essential element allowing these particles to combine: the *electric force*. The electric force is well documented by the Standard Model and its strength and characteristics are well known to scientists and engineers who use that information extensively throughout science and industry. While the strength and attributes of the force are well understood, the source of the force is only theorized, and that theory is not particularly believable.

The Electric Force

Eighteenth century scientists and others including Ben Franklin were familiar with the rudiments of electricity being the motion of electric charges, which were labeled positive and negative. French scientist Charles-Augustin de Coulomb established in the late 1700's that charged objects attract and repel each other. The connection between electricity and magnetism was discovered in the early 1800s by Michael Faraday. Faraday formulated the notion of magnetic and electric fields, and that particles passing through those fields would experience a magnetic and electric *force*.[3]

The source of the force was initially mysterious and was thought to be an inherent quality in the charged object. Quantum theory now theorizes that force arises from the exchange of virtual particles called bosons, discussed below. The Tor Model ascribes the electric force to be inherent in the fields of Tors. Whether a particle has a positive or negative charge depends on the dominance of left or right spinning Tors it contains. A Tryk with a dominance of left-spinning Tors (*shown in Fig. 3.2a*) possesses a *positive charge.* A dominance in right-spinning Tors (*shown in Fig. 3.2b*) has a *negative charge.*

The respective fields create an attractive *force* for Tryks with opposite-charges and a repulsive *force* for Tryks with like-charges, as shown in Figure 4.1.

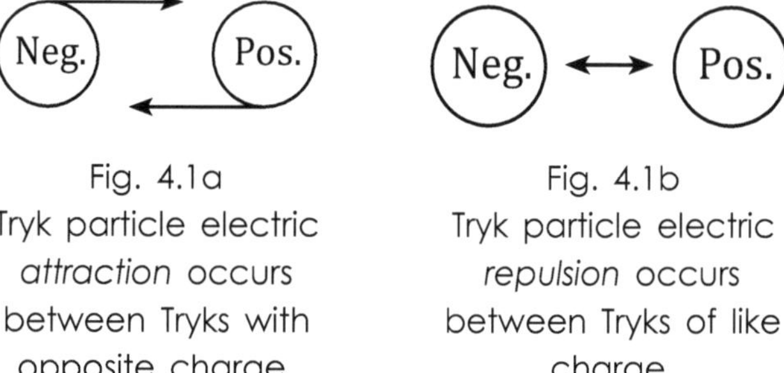

Fig. 4.1a
Tryk particle electric *attraction* occurs between Tryks with opposite charge.

Fig. 4.1b
Tryk particle electric *repulsion* occurs between Tryks of like charge.

The Source of Electric Force According to the Standard Model

In the Standard Model the electric force is conveyed by the respective particles exchanging virtual *photons.*[4] A photon exchange between two like charges (*two protons*) creates a repulsive force, and between two unlike charges (*electron and proton*) an attractive force.

It is difficult to fathom how the exchange of a particle of any kind can create both an attractive and repulsive force. The theory is that when an electron or quark emits a photon, it changes that particle's velocity, and when the other particle absorbs the photon it changes that particle's velocity, just as if there had been a force between the two particles.[5] It is especially difficult to believe the exchange creates a consistent force when the exchange particle in question is a *virtual* particle. The following observations make the Standard Model's force theory by virtual-particle exchange difficult to believe.

- A virtual particle is one that theoretically pops in and out of existence. How can there be a consistent and smooth *force* while relying on something that is there one moment and gone the next, no matter how quickly the exchanges take place?
- Virtual particles are hypothetical, never directly observed, and are only believed to be real because there is no other reasonable explanation for such things as the Casmir Affect (*two metal plates being pushed together by unseen forces*)[6], and because mathematically if the virtual particle's presence is not included in the calculations the answers are wrong.[7]
- Virtual-particle exchange often results in calculations that include *infinities* and the only way to make sense of the calculations is to simply ignore or remove them. Removing them has been given the innocuous name of *renormalization*; a procedure that is admittedly dubious mathematically.[8] None the less, the interaction between photons and matter is summarized in a body of equations known as Quantum Electrodynamics (QED), which has proven very accurate and is used throughout science and industry. Again, the Standard Model has found a way to make the math work, but it doesn't mean the underlying theory involving the exchange of virtual bosons is correct.
- Of the four virtual-exchange particles, the Graviton, theorized to convey the gravitational force, has never

been detected.[9] The graviton's absence brings doubt to the entire scheme of boson exchange being the source of force. Even if gravitons are miraculously found, the interaction between matter and gravitons produces infinities that cannot be renormalized (*think done away with*), so at present graviton exchange is not even a viable theory.[10]

- Kenneth Ford and others tell us that every time an electron absorbs or emits a photon, entire new particles are created and the original particles are destroyed.[11] If the original emitting electron is turned back into a knot of Penergy and from that knot a whole new electron and photon are created, and this happens every time a photon is absorbed and emitted, it seems like a *very* laborious, inefficient, and unreasonable way for nature to work.
- The Tor Model posits force being conveyed through the fields of particles, and scientists have noted that fields take time to exert their influence. Physicist George Musser notes that a time lag seems odd if forces are leaping directly from one object to another [*as in particle exchange*] but is perfectly natural if an impulse must make its way through a medium.[12] In other words, one would expect a force conveyed through a boson exchange to be initiated smoothly, without a time delay, but evidently that is not what scientists observe. The noted time delay makes sense if forces are conveyed through fields.
- Though the electric charge field gets weaker with distance, its influence is supposedly without boundary. This makes an electron on earth having a very small influence on an electron on Alpha Centauri, four light years away. If this is so, it would mean that the electron is constantly exchanging virtual photons with that

electron.[13] It would also mean the electron was making the same constant exchange of virtual photons with every other charged particle in the space in between! Again, that makes this scheme laborious and inefficient, making it difficult to believe nature works that way.

- Theoretically, for virtual particles to come into existence they borrow energy from what is called the *vacuum of space*. If virtual particles are popping in and out of existence constantly to affect forces, it would require a great deal of energy to bring all of them constantly in and out of existence, even momentarily. According to Tim James in his book *Fundamental,* quantum theory can predict how much energy density there should be in the vacuum of space by adding all the virtual particles together. Doing so produces a theoretical energy value of 10^{105} Joules (*think ounces of energy*) per cubic centimeter. Examining space, however, produces a value of only 10^{-15} Joules per cubic centimeter. Comparing what the energy level should be to what is measured, scientists find the theoretical value to be 10^{120} times larger than the actual value.[14] Michio Kaku characterizes this as the largest mismatch in the entire history of science.[15] Virtual particles might exist, but apparently not in the numbers to support them being the source of all the forces.

According to quantum physics, the vacuum of space is awash in virtual particles constantly popping in and out of existence.[16] The Tor Model recognizes the existence of virtual particles but not to the degree theorized by quantum physics. In the Tor Model, virtual particles only come into existence when a fluctuation in the interstellar medium creates a small knot of Penergy. The knot of Penergy provides a momentary rise in the Penergy density sufficient to bring particles into existence.

But because the knot begins immediately dissipating back into the surrounding thin-density Penergy, the failing knot is not strong enough to sustain itself, so the particle(s) disappear back into the medium before being fully formed. Consequently, virtual particles are not popping in and out of existence everywhere, only where fluctuations in the Penergy medium are prominent such as near blackholes and other very dense masses, or within the environment of a planet's atmosphere such as our own.

Charged particles may absorb and emit photons, even virtual photons that affect each other's momentum, but that fact by itself does not necessarily create an electric *force*. The electric force is important since it is instrumental in the evolution of future levels of matter. We will look closer at the Tor Model's view of the electric force when examining the configuration of the up quark. First let's begin our examination of subatomic particle configurations by looking at the prospect of a small but necessary and possibly unknown particle — the connector.

Subatomic Particle Configurations

Tryk Connectors

Tor-Chains and Tryk-Chains are easily assembled and probably accounted for much of the early composite particle creations, growing to hundreds and possibly thousands of particles in length. Such lengthy chains would not necessarily have been conducive to creating the next level of complex particles. They could, however, account for much of the dark matter present in the universe. Tor-Chains are composed of only one kind of particle in an aligned chain, so it could grow to be very long, but for Tryk-Chains it may have been a different story.

As previously observed, it was the strong Tor field that encouraged a nose to tail coupling creating Tor-Chains, and that same coupling capacity would create Tryk-Chains, as well. However, with the addition of an opposite-spinning Tor circling the chain to create the Tryk, the overall spin of the Tryk would likely wobble a bit, as illustrated in Figure 4.2.

Such a wobble would possibly weaken the field-connection strength of the combined Tor-Chains. As the Tryk-Chain links grew longer, the magnetic field strength of the Tors holding the Tryks together would have grown weaker. The wobble may not have been a problem for very short Tryk-Chains but may have made the creation of lengthy Tryk-Chains problematic. That wobble might be the source of the small precession value found in all electrons.

This limitation to Tryk-Chain length could have been overcome if there had evolved another independent particle that could serve as a flexible connector (⬡) between the Tryks. In this way the Tryk-Chain connection would not be completely dependent on the magnetic force between Tors, allowing for stable Tryk-Chains of any length.

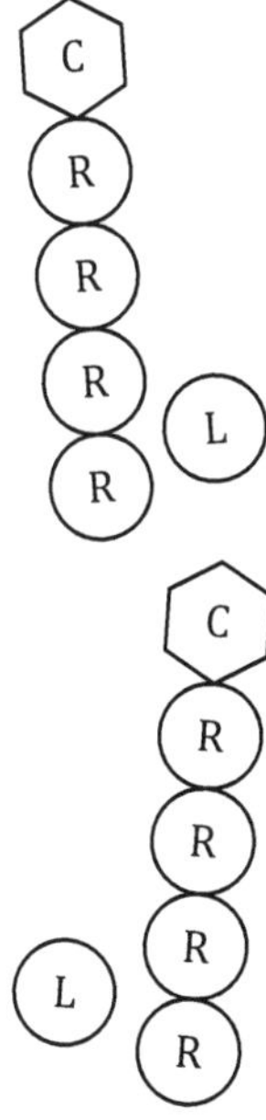

Fig. 4.2 Multiple rotating Tors could cause Tryks to wobble.

Such connectors would likely be very small and neutral in charge. They might exist without our even knowing it. Being small, neutral, and likely near massless, in the collision debris calculations in colliders and accelerators they could easily be mistaken for neutrinos.

Neutrino Connectors

Neutrinos are one of the most abundant particles in the universe. They have neutral electric charge, are nearly massless, and interact with other matter very little.

In the Tor Model, Neutrinos may serve as a connector particle for electrons and positrons. Neutrinos only seem to appear in conjunction with an emission of an electron or positron due to particle decay or during the fusion process at the heart of a star. Every time an electron or positron is spit out, so is a neutrino. This close tie suggests a possible physical connection between the two particles.

The neutrino may be a *Majorana* — a particle that is its own anti-particle.[17] Experiments are currently underway to determine this. If the neutrino is its own anti-particle, it would have a unique configuration. As such, it may serve as a universal connector allowing for the connection between all sorts of Tryk-Chains and other particles.

To propose a possible configuration for the neutrino we will have to know more about its behavior, attributes, and relationship to electrons and positrons.

Quark Configurations

According to the Standard Model, protons and neutrons are composite particles comprised of three quarks and a sea of virtual *gluons* that hold the quarks together. Quarks and gluons have not been isolated or observed directly but are believed to exist based on experimental evidence.[18] Quarks are believed to be elemental, are never found as a single unit, and are always found paired together in twos or threes. They are theorized to carry electric charge but only in 1/3rd or 2/3rds of the whole 3/3rds charge value of the electron.[19] The theory posits no explanation for why quarks have fractional charges.

In the Tor Model, quarks are not elemental but are comprised of Tryks, and their fractional charges can easily be constructed from those Tryks. Given that Tors come in matter and anti-matter pairs that have a natural attraction/repulsion for each other, in the Tor Model that attraction/repulsion relationship is the source of the *electric force,* as illustrated in Figure 4.1.

It would mean that *charge* is not an inherent characteristic of electrons, protons, and quarks per se, but an inherent characteristic of the Tor field. That would make the Tor field arguably responsible for both the magnetic and electric forces. Scientists have long considered the magnetic and electric forces to be related, so it makes sense to have both forces stem from a natural attribute of the most elemental particle, the Tor field.

We have now come across three circumstances involving an attractive/repulsive force. Let's take a moment to examine this phenomenon for clarity. The discussion will be placed in a table for ease of comparison.

Attraction Between Like-Spinning Tors.
Left and/or right-spinning Tors relate to each other through their fields. When the fields of two or more like-spinning Tors are aligned nose to tail the Tors are attracted to each other creating Tor-Chains. If aligned in any other configuration, the fields will repulse each other.

Matter/Anti-Matter Mirror-image particles.

Matter and anti-matter are mirror-image particles comprised of combinations of left and right-spinning Tors. A predominance of one Tor-spin we call matter and its mirror-image with the opposite spin we call anti-matter. Matter and Anti-Matter attract each other, and two like-matter particles repulse each other. The source of the attractive/repulsive force is inherent in the fields of the underlying Tor particles.

Composite non-mirror-image particles.

The attraction between two non-mirror-imaged particles is called the Electric Force. It is simply the same attractive & repulsive dynamic as Matter and Anti-Matter, except that we ascribe them the trait of negative and positive *Charge*. What we label as opposite charges, however, is simply particles with an abundance of right and left-spinning Tors. Just like matter and anti-matter, it is the relationship between the underlying opposite-spinning, Tor-fields that creates the attractive force.

In summary, the matter/anti-matter, electric, and magnetic forces are all derived from the same source — the attractive/repulsive relationship between the underlying Tor fields.

Up Quark Charge Values

As mentioned earlier, the strength of the electric force is measured in *charge value* using the electron as a basis. It has a charge value of -1, since it is ascribed to have a negative charge, while its anti-matter particle the positron is ascribed the charge of +1.

As observed, quarks possess Electric charge but only in amounts $1/3^{rd}$ and $2/3^{rds}$ that of the electron. For the sake of demonstrating how these partial Electric charges are possible at the Tryk level, let's give each individual Tor within the Tryk a $1/9^{th}$ charge value. Accordingly, right-spinning (*negatively charged*) Tors would have a $-1/9^{th}$ charge value, and a left-spinning (*positively charged*) Tor would have a $+1/9^{th}$ charge value. These particular values are ascribed for demonstration purposes only.

With these values and terms, we can construct all sorts of particles with all sorts of electric charge values both positive and negative. Because Tryks contain both positive and negative Tors, each pair of which would add to zero, we only need to calculate the charge value based on the *net number* of charges.

To get the net number of charges for each Tryk we add up all the charges. As shown in Figure 4.3a, a Positive Tryk would have a gross charge value of $+4/9^{ths}$ for the four positive (*left spinning*) Tors, and $-1/9^{th}$ for the one negative (*right spinning*) Tor, giving it a net charge of $+3/9^{ths}$, or a $+1/3^{rd}$ overall charge value.

As shown in Figure 4.3b, a Negative Tryk would have a gross charge value of $-4/9^{ths}$ and $+1/9^{th}$, giving it a net charge of $-3/9^{ths}$, or a $-1/3^{rd}$ overall charge value. In summary, the Electric charge value of any particle is determined by the net differential in the positive and negative Tor count making up the particle.

The values given above are arbitrary and are used only for sake of example. Each Tor could have been assigned a $1/90^{th}$ charge value, which would work out the same as long as the other values were in the same proportion. It is not the goal to prescribe a specific charge value for the Tor but only show how charge values *might* be distributed between the applicable particles.

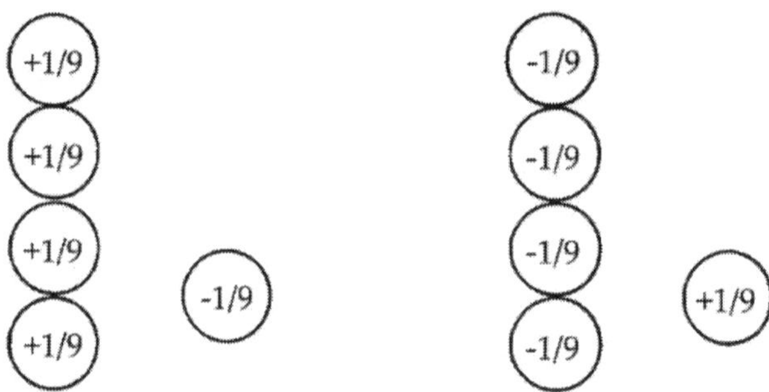

Fig. 4.3a. Charge values of a Positive Tryk: net +3/9ths or a+1/3rd overall charge.

Fig. 4.3b. Charge values of a Negative Tryk: net –3/9ths or a –1/3rd overall charge.

Figure 4.3 Charge values of Tors added together to give a *net charge* value of a Tryk.

Deduction #10: Given the arguments for the source of electric force not being due to virtual boson exchange; and given the simplicity of that source being the natural attraction/repulsion between early elementary Tor particles; we can conclude that the source of the electric force is a differential in the number of positive and negative Tors making up a particle: an abundance of positive Tors giving the particle a positive charge, and vice versa.

Prediction: An enterprising young physicist will someday extrapolate the mathematics of quantum field theory into a theory of the magnetic and electric forces conveyed through the fields of early elementary particles.

Simple Quark Configuration

Given these Tryk charge values we can now easily create a composite particle with an up quark charge value of

+2/3rds by joining two positive Tryks into a single chain as illustrated in Figure 4.4a. The anti-particle of this simple illustration of an up quark would have the mirror image configuration and carry a –2/3rds charge, as illustrated in Figure 4.4b.

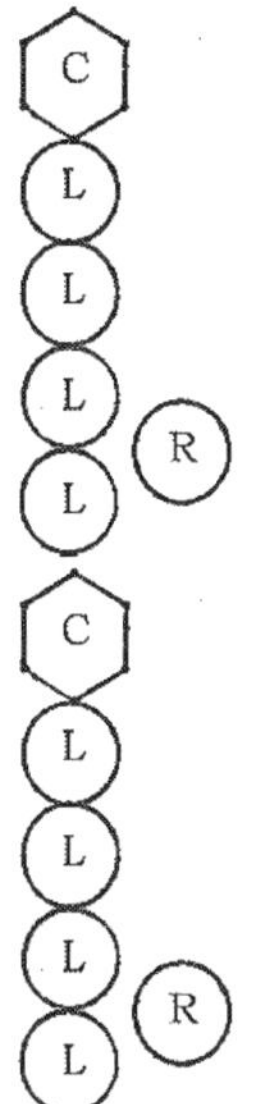

Fig. 4.4a. Two Positive Tryks join to create a +2/3rds charge quark.

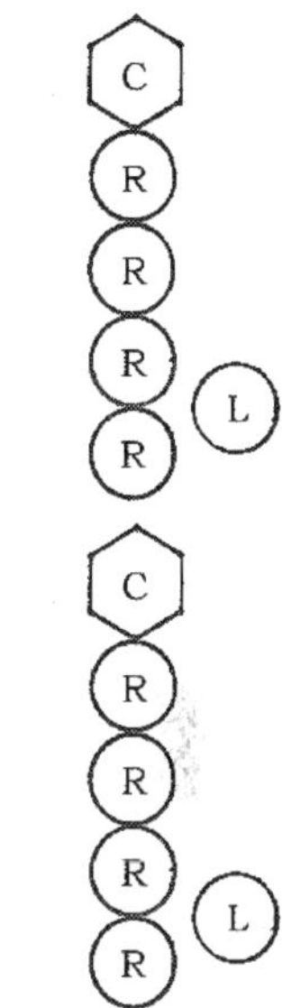

Fig. 4.4b. Two Negative Tryks join to create a -2/3rds charge anti-quark.

Complex Quark Configuration

A quark configuration might be as simple as Tryks hinged together, but it is more likely that out of the frenzy came much more complex particles. A down quark for example might be made up of a chain of electrons, positrons, neutrinos, universal connectors, and positive and negative Tryks. An example is illustrated in Figure 4.5.

This configuration would have a net -1/3rd charge value, which is the exact charge value of the down quark. Should it spit out the electron/neutrino couplet at its corner, it would then have a +2/3rds charge value, becoming an up quark, changing a neutron to the proton. Likewise, should an up quark spit out the positron/neutrino couplet at its corner, it would then have a -1/3rd charge value, becoming a down quark, changing a proton to a neutron. The ease at which particles decay into other particles in this fashion makes the argument for composite particles compelling. This down quark configuration is not definitive and is only one of many possible configurations for a quark made from Tryks.

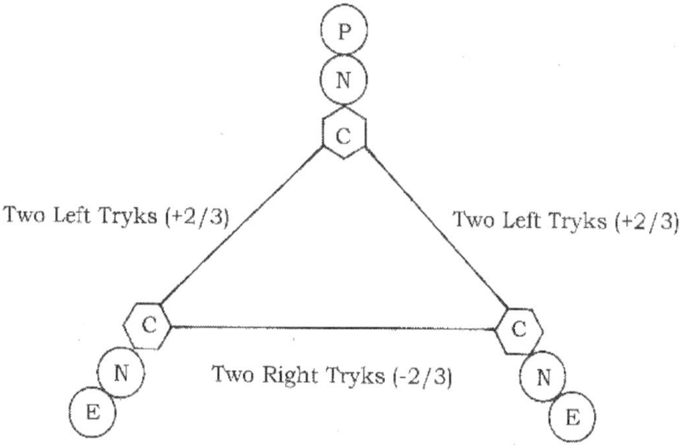

Fig. 4.5 Possible Complex Down Quark: (E) Electrons, (P) Positron, (N) neutrinos, (C) Universal Connectors, and positive and negative Tryk pairs.

Although many Tryk combinations would have been created, natural selection chose our up quark with a +2/3rds charge and down quark with a -1/3rd charge to dominate. These quarks would have become dominant due to their need to be in threes to be stable, and to combine their partial charges to add up to the +1 charge (*+2/3, +2/3, -1/3*) of a proton, equaling perfectly and offsetting the -1 charge of the electron. Those configurations eventually allowed the hydrogen atom to form.

Quark Evolution

That sequence was a lot to take in so allow me to rephrase it for clarity. In the Penergy medium frenzy, many complex quarks comprised of various particle-chain combinations, with various charge values, would have been created. But not until a stable, three-quark combination, with a +1 net charge evolved did natural selection allow the up and down quark configurations to become dominant.

But how did the universe know it needed a +1, positively charged particle? It was not a conscious need but a need for cosmic evolution to progress. The CAGI that was briefly described in Chapter Two compels matter to combine until it creates a new level of matter with its own capacity to combine. Time is of no concern. The CAGI will take as long as necessary and create as many trial combinations as necessary for a combination to come about that qualifies as, or is a sub-stage to, the next stable Level.

In this case, quark-like combinations and electron-like combinations would be created over and over until configurations capable of combining into a larger, stable particle came about. Since the electric force was now available to bring particles together, it makes sense that the primary attribute driving the evolution of these particles was their electric charge value. Once the stable electron with a -1 negative charge became dominant, it was only a matter of time before a Tryk combination came about to precisely off-set that charge value and create the proton.

Meson Configuration

A meson is a two-quark combination comprised of a quark and an anti-quark. There are many combinations of a quark and anti-quark having different spin signatures that preclude annihilation: an up quark and anti-down quark (*net charge +1*); or a down quark and an anti-up quark (*net charge -1*). A mirror-image quark pair that won't annihilate can also combine to make a meson if the parts come together fragmented so that no mirror-image elements are ever present.

In the Standard Model, the mirror-image quarks contain a different *color charge,* so they don't annihilate. The color charge is part of the Standard Model's scheme for particles and forces within the nucleus. A meson can also be comprised of a quark and anti-quark from different generations. Mesons are, however, unstable and suffer an immediate forced decay when created in a shower of collider particle debris.

Electron Configuration

The lepton group is comprised of electrons and neutrinos. According to the Standard Model, the electron is an elementary particle without constituent parts. Frank Wilczek reports in his recent book, *Fundamentals — Ten Keys to Reality,* that electrons can be fractured into smaller particles with partial charges through the *fractional quantum Hall effect.* But these are not your everyday electrons. They are created under precise laboratory conditions and are called quasiparticles.[20] It appears electrons could have an inner structure, even if it's only manifested under a very limited and extreme experimental environment.

In the Tor Model, the electron is a composite particle made up of Tryks. The key element in the configuration of an electron is that it carries a whole 9/9^{ths} negative-charge value, which means that using the values we established already it has a differential of nine negative Tors. Three negative Tryks in a chain would fulfill the requirement as illustrated in Figure 4.7a. The positron (*anti-electron*) would of course have the mirror-image configuration as illustrated in

Figure 4.7b. Other Tryk and particle combinations comprising an electron are of course possible.

Fig. 4.7a. An Electron could be made from a chain of three negative (**R**ight) **T**ryks.

Fig. 4.7b. A Positron could be made from a chain of three positive (**L**eft) **T**ryks.

The electron illustrated above is made from a *chain* of Tryks. One might consider creating an electron from three unconnected, independent Tryks. That would seem equally feasible, but it is not possible due to the *Pauli Exclusion Principle.* That principle says that no two like fermions (*electrons or quarks*) can occupy the same state, meaning carry on the same function, together. We need not go into the principle further but know that three Tryks in a chain is more likely to comprise an electron than three single, independent Tryks.

In the chain configuration, the three, single, rotating Tors can co-exist because they are not rotating in the same plane and therefore are not in the same state. The Tryks being in a chain is probably the reason the electron is thought to be elementary since the chain of Tors comprising the Tryk would have a strong coupling capacity and be very difficult to break apart.

Even without the existence of a connector, three Tryks may have been able to join nose to tail to create a strong chain despite the repulsive aspect of the electric force. Firstly, we have concluded that due to the very strong field strength of the Tor, in the early universe the magnetic force could reasonably overcome the electric force.

Secondly, we have also observed this to be the case in experiments involving two otherwise repulsive electrons coming together nose to tail to create a combination called a Cooper Pair.[21] This phenomenon came to light when scientists were exploring theories of superconductivity (*the flow of electricity without resistance*). Again, it only happens in extreme laboratory conditions. We know, however, that the strength of forces bringing particles together can change under different energy conditions, and the early universe certainly provided different energy conditions.

Boson Configuration

According to the Standard Model, bosons are virtual particles that are theorized to mediate (*think facilitate*) the absorption, emission, creation, and decay of Sub-A's. The group is comprised of photons, gluons, gravitons, W and Z Particles. Being virtual means they exist only momentarily, flitting in and out of existence.

The Tor Model recognizes their existence but believes their importance in our universe is uncertain. The Tor Model has an alternative explanation for the source of force and how particles interact, so the possible makeup of these particles will not be addressed. They will be discussed further in Chapter Five under the topic of *Forces.* Photons are known to also be stable particles, so let's examine how they may have evolved.

Photon Configuration

Photons are widely known as *light.* They are the particles that are absorbed and emitted by matter at specific wavelengths, which then enter our eyes, creating impulses that are interpreted by our brains, giving us vision. Photons do not naturally interact with each other. Except when passing through a clear medium, such as air, glass, or water, they travel in a straight line and at light speed, roughly 3.0×10^8 meters/second.

Photons are never at rest, so there is no way to measure their rest mass, but experiments indicate they are massless. They can be *absorbed* and *emitted* by charged particles, e.g., quarks and electrons, but otherwise exist only in constant motion within the Penergy medium. When absorbed, their energy is transferred to the particle, boosting its *kinetic energy* (*energy associated with motion*).

Photons can be very high energy and can decay (*think change*) into most any kind of matter-antimatter pair. They can do this while remaining massless, making their composition extraordinary. They may or may not be comprised of Tors. If not, it would mean the early Penergy also spun itself into an entirely different kind of stable particle that is peculiar to the photon. We will proceed with the notion photons are comprised of Tors, but other options may be equally credible.

To be charge neutral, photons must have an equal number of left and right-spinning Tors. Having little or no mass means their Tor spin is in perfect, or near perfect, balance to have little if any gyroscopic effect. Having no mass, photons do not distort the surrounding Penergy medium, thus have no gravitational effect on other particles.

All these characteristics might be accounted for if the photon is comprised of a loop of both Left and Right-Tor-Chains with equal length differences, as illustrated in Figure 4.8a. The loop length and number of chains are unknown, so the lengths appearing in 4.8b are merely for example. This multi-stranded, loop configuration is easy to construct being comprised of simple Tor-Chains. It can be easily assembled from a knot of Penergy produced by an annihilation, and easily broken apart and reconfigured to produce any matter-antimatter pair.

This design also satisfies the fact that the mirror image of a photon looks the same as the original, making them their own anti-particle.[22] We will look closer at the photon in Chapter Five under the heading *The Electromagnetic Field.*

Fig. 4.8a End view showing number of Tors in each chain.

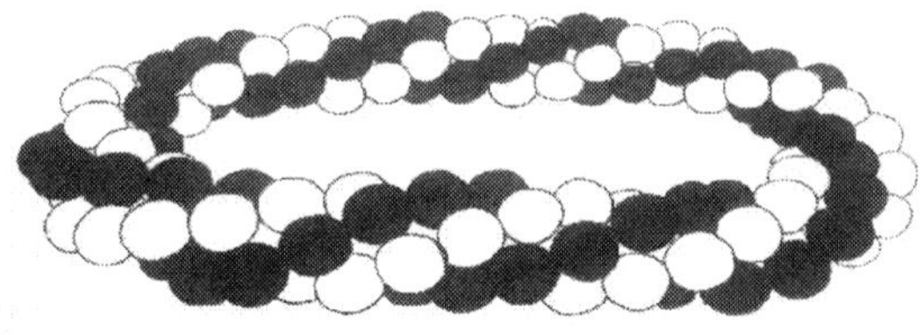

Fig. 4.8b Possible photon design made up of four Tor-Chains of different lengths.

Again, these configurations of quarks, electrons, neutrinos, and photons are put forth only as an example of how these composite particles might be configured. The only attribute that is certain is that they are all comprised of a single early elementary particle like the Tor.

The CAGI and natural selection can produce some very complex pieces of matter with some very interesting emergent characteristics. While we dig deeper into the secrets of nature, we should always be ready to take a step back and recognize such qualities. It is easy to miss the forest through the trees, in this case the emergent qualities through the constituent parts.

This completes our discussion of the evolution and creation of the individual Sub-A's. Before examining how these composite particles came together to create the next level of matter, we will examine in Chapter Five some of the attributes and conditions relating to Sub-A's.

* * *

A premise of the Tor Model is that our subatomic particles and forces did not inexplicably pop into existence but are a consequence of particles and forces that evolved from early elementary particles. In this chapter we examined the source of electric charge and found it more likely to be derived from the fields of Tors, and that *charge* is simply the differential between the number of matter and anti-matter Tors making up a particle. Recognizing we cannot presently know the exact makeup of our Sub-A's, we put together pictures of each type of particle to show how it *might* be configured.

Chapter Five

Building Block Characteristics

Our universe now contains the third level of building blocks, the Sub-A's, but the hot and frenzied environment is too energetic to allow those particles to combine for now. They are in a hot, *plasma* state that only exists in very high, energetic conditions. While we await the universe to further expand and cool, let's look closer at some of the characteristics of the particles thus far created and some of the conditions that flowed from their evolution.

The essential emergent quality inherent in Sub-A's is the size, capacity, and cumulative effect of their fields. Identifying, quantifying, and learning to utilize those fields has powered our industrial and scientific advancement for the past two hundred years.

A field is an area of space immediately surrounding a particle that has a measurable effect on other particles passing through that space. A particle's field plays a significant role in determining how particles interact, combine, and at what strength. Fields were once thought to be a product of particles, but now the opposite is true; current theory has particles being derived from fields. Let's take a closer look at this important topic.

Particle Fields

A field for our purposes is an area of space containing a common, measurable characteristic. The field can be large, such as an entire weather front, wherein each point in space has a measurable temperature. The overall area so measured would be considered a *temperature field.* A field can be small, such as the field surrounding a particle possessing an electric charge. At each point in that space there is a measurable charge value and the overall space so measured would be a *charge field,* a.k.a., *electric field.*[1]

Particle fields do not have precise boundaries since the measurable values gradually dissipate with distance from the particle. They do have a practical boundary, which is the margin at which the field has an insignificant influence on other particles.

Opinions differ as to how fields are created. Do fields create particles, or do particles create fields? In Quantum Field Theory (QFT) fields create particles. Those fields are ubiquitous throughout the universe and for every particle there is a specific field. QFT says particles have no size, take up no space, and are only manifestations of a field.[2] Particles are also said to be excited states of their own fields.[3] A quark is simply an excited (*think energetic*) state of the quark field.

QFT and the Tor Model have similarities in their view of the universe. QFT recognizes particles and fields are associated, but it is the fields that are ubiquitous throughout the universe and from which everything else is made.[4] In the Tor Model, Penergy is ubiquitous throughout the universe and from which everything else is made.

In QFT, the ubiquitous fields are the elements that play the role of a communications network between particles.[5] In the Tor Model, Penergy is the element that plays the role of a communications network between particles. Everything discussed in the prior chapters suggests Penergy is real. There is no evidence ubiquitous fields are real, outside of mathematical equations.

QFT is another mathematical edifice that describes a reality difficult to believe. Physicist George Musser tells us physicists struggle to understand what quantum field theory is telling them about the world.[6] Chemical Physicist Michael Munowitz says that, 'a quantized field at first blush is a monstrosity, a mathematical absurdity that makes no sense at all.'[7]

Physicist Sean Carroll tells us, 'The Core Theory [*a.k.a. Quantum Field Theory*] is not the most elegant concoction that has ever been dreamed up in the mind of a physicist, but it's been spectacularly successful at accounting for every experiment ever performed in a laboratory here on earth.'[8] Quantum math is torturous, but it works. But that does not mean QFT's *interpretation* of how that math reflects on reality is correct.

QFT might be mathematically accurate, but it seems more reasonable that it is the particle that creates the field. In that case the field is simply a local disturbance in the Penergy medium due to the dynamics of a particle's vibrating presence. We need to understand fields to build our universe, but we don't necessarily need to resolve which came first, the field or the particle. But again, it is fun to speculate on a different point of view of what a field might actually be.

Speculations on the Source and Attributes of Fields

Scientists have observed that particles such as protons and neutrons vibrate with their own specific frequency.[9] In the Tor Model, given the presence of the fluid-like, thin

Penergy density that comprises our interspatial medium, it makes sense that a vibrating particle would affect the surrounding Penergy medium. In doing so it would create a disturbance that pulsates in waves with typical wave-like features, such as wavelength, intensity, and frequency.

This point may be at the heart of the issue as to how a particle can exhibit both point-like and wave-like characteristics. The matter is commonly referred to as the wave-particle duality issue and is a core feature at the foundation of Quantum Theory. We will examine the duality issue in detail in Chapter Seven.

Particles affect the Penergy medium in proportion to the disturbance they make, making the field size generally in proportion to their makeup, spin characteristics, and energy. Tor particle fields exhibit a natural attraction when the Tors are aligned nose to tail, creating Tor chains and a magnetic field. A magnetic field is much bigger than its source, just as the electric field is much bigger than the electron. These observations imply that the field of even a single particle is much bigger than the particle itself, surrounding the particle like a cocoon or bubble. As previously observed, the cocoon does not have a specific boundary; the field gradually fades into a practical boundary.

Physicists have espoused the idea of particles creating wave-like fields. Physicist Michio Kaku tells us that the photon particle creates both an electric and magnetic field surrounding it, and the fields are shaped like waves and obey Maxwell's equations (*summarizing the electric and magnetic relationship*).[10] That observation suggests that it is the field that is doing the waving, though I'm unsure whether Dr. Kaku was intending to make that specific point.

Speculating further, if the particle was standing still and not being influenced by its environment, the surrounding bubble-like field would be fairly uniform in its distance from the particle. A moving particle in its customary environment, however, must deal with all the other particles, fields, and elements cluttering the environment, causing it to rapidly move, dodge, and respond, distorting and constantly reshaping its field.

We can reason that at any moment the probability of finding the particle within any specific area of its field is in proportion to the relative strength of the field at that point, which is how quantum mechanics work.

Quantum mechanics is a mathematical formulation that provides an accurate probability of an outcome in particle experiments. We can only know a probability rather than a certainty for some experimental outcomes due to the vagaries of finding a particle at a specific place, at a specific time, within its field. We will go into this idea further and discuss a particle's individual field at greater length in Chapter Seven.

We have observed from electric fields that the presence of many of the same particle have the cumulative strength of all the particles involved. As different particles combine, their fields combine forming a new field holding the characteristics of the new combination of particles, e.g., a proton's field is much different than the fields of the individual quarks making up the proton. As particle/bodies combine and gain more mass, their fields, though more complex, are not necessarily proportionally larger.

As the particle/body scales up in mass, the movement of the particle/body within its field is gradually less determined by the subtleties within its environment. Once it reaches a scale where the particle is seldom making responsive moves within its field, it leaves the quantum world and joins the macro-world. In the macro-world the attributes of the particle/body can be measured with certainty, no longer needing Quantum Mechanics, though it remains quite accurate if used. Fields remain an attribute of larger bodies, including humans who refer to their personal fields as auras.

Quantum Theory recognizes particles having attributes, but there is a controversy as to where some of those attributes reside and when they come into existence. One would think that a particle has a certain attribute, or it doesn't. And it should not matter whether that attribute has a fixed value such as charge, or a variable value such as its direction of spin. Apparently in quantum mechanics that is not the case. Philip Ball in his book *Beyond Weird,* tells us that quantum mechanics does not recognize particle variables that can be assigned in advance even though they appear to acquire those values randomly through the act of measurement.[11]

Consequently, those variable characteristics do not appear in quantum theory's mathematical descriptions of the particles.[12] In other words, despite knowing that electrons are either spin-up or spin-down, its spin direction *does not exist* until it is measured and determined. Those variables that particles can possess were deemed hidden from view and became known as *hidden variables.*[13] Whether variables capable of being pre-set exist prior to measurement is an open issue and is at the heart of a phenomena called *entanglement,* which is another core element supporting Quantum Theory and will be discussed at length in Chapter Seven.

In the Tor Model, particles have many attributes or variables, all of which are real, whether hidden or not. Some of these attributes are represented in their fields, as listed below. There may be more.

- Since all particles vibrate at their own peculiar intensity and frequency that is reflected in the particle's field, if one could *read* the field, one could identify the particle by those attributes.

- A photon's state includes some value for its polarization.[14] Polarization is the direction in which the photon is determined to be doing its waving. Since that which is waving is the photon's field, then its polarization must be said to be attribute of its field.

- Since a field has a wave-like attribute, and the frequency of the wave is a function of the particle's energy, one can say that the field is also a reflection of the particle's energy.
- A particle's response to another particle's field is what is said to be a force. The source of force thought to be inherent in a particle is actually inherent in, and a function of, the particle's field.

Some believe the particle's field even allows for communication. Lynne McTaggart tells us in her book *The Field,* 'These subatomic waves or particles not only know about each other, but also are highly interlinked by bands of common electromagnetic fields, so that they can communicate together.'[15]

Notwithstanding the approach taken by Quantum Theory, it would seem those variable characteristics are real and a part of the particle, whether measured or not. There may be more to particle fields than the current mathematical model is telling us. The subject of fields warrants additional investigation. Further evidence for local particle fields will be discussed in Chapter Seven, and how fields play into our concept of energy will be discussed in Chapter Eight.

> Deduction #11: Given the cumulative evidence for the existence of the Penergy medium; and given that the Penergy medium can serve in the same capacity supporting particle fields as QFT's universal fields; and given the absence of evidence for the existence of universal fields except mathematically; we can conclude that the source of particle fields is more likely a disturbance in the Penergy medium caused by the spin characteristics of individual particles. Additional support for this deduction will be revealed in subsequent observations and deductions.

Of the many fields that can be created by particles and particle combinations, three fields, in addition to the gravitational field, have a big influence on our everyday lives: the electric, magnetic, and electromagnetic fields.

The Electric Field

Recall that in the Tor Model, charge is the trait ascribed to the natural attraction/repulsion between individual left and right spinning Tor-particles. A charged particle is one comprised of a differential in the number of Left-Tors and Right-Tors. The strength of the electric charge is in proportion to the size of that differential.

Charged particles such as protons and electrons are naturally surrounded by their own field containing all their attributes, including the effects of their charge differential. If one lines up a row(s) of protons on one side and a row(s) of electrons opposite them, with their respective fields overlapping, it creates an attractive electric differential that will induce electrons to move across the combined fields creating electricity. Electricity of course has multiple uses and is very important to our comfort, technology, and existence.

Given a reasonable distance between them, the more protons and electrons that are involved, the stronger the combined field. The strength of the electric differential between the positively and negatively charged rows is measured in *volts.* An electric charge differential can also be *induced* in a wire by the movement of an adjacent magnetic field, which is how a generator works. There will be more on this subject in Chapter Six under the heading of *Electromagnetic Force.*

The Magnetic Field

In the Tor Model, as noted earlier Tors combine by lining up nose to tail creating chains with spin axes all in line. The

Tor-Chain creates a surrounding field, the effect of which causes other nearby Tors to align their spin axes with that of the Tor-Chain. This spin-alignment force carries over to Tryks, electrons, and atoms, as well.

Consequently, bound particles such as electrons can have their spin axes aligned, and the alignment of multiple-particle spin-axes creates a recognizable magnetic field. The field is attractive to another magnetic field when the axes are aligned nose to tail, and repulsive when aligned otherwise, especially nose to nose or tail to tail. Again, the more particles involved in the alignment process, the greater the size and strength of the magnetic field.

Substances with their atomic spin axes aligned creating a magnetic field are called magnets and they can be created in several ways. A substance having permanent alignment of the spin axes of its atoms is a permanent magnet and is found naturally in lodestone (*magnetite*). A magnetic field is also induced when a current of electricity is moving through a wire, creating an electromagnet. A magnetic field is created by a changing electric field and vice versa, which is the technology used in the creation of electric motors and generators.

The Electromagnetic (EM) Field

The photon is believed to carry both the electric and magnetic fields and to be the source of the combined *electromagnetic field*. Its configuration is unlike any other particle. Photons may be configured in a loop from parallel, opposite-spinning Tor-Chains as illustrated in Figure 4.8. Photons spin on their long axis as the loop twirls through space. In a beam of photons with their spins aligned, the angular momentum of the individual photons adds up and the beam imparts a measurable torque or twisting force.[17] This suggests that photons do indeed spin on their long axis. The loop traveling on its long axis assumes the shape of a long Tor-Chain that naturally creates a *magnetic field,* as illustrated in Figure 5.1.

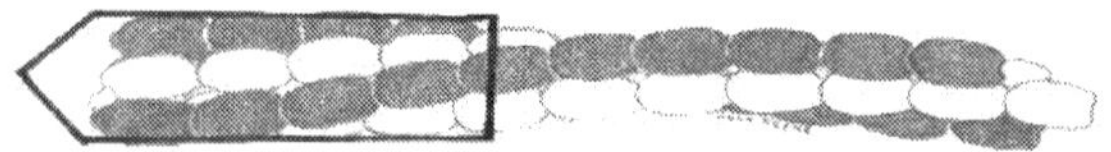

Figure 5.1 A twirling photon loop assumes the shape of a, long Tor-Chain, naturally creating a magnetic field.

Because the photon spins on its long axis, the rotating Left-Tor and Right-Tor Chains reflect a differential in Tor counts, mimicking an *electric field* as illustrated in Figure 5.2.

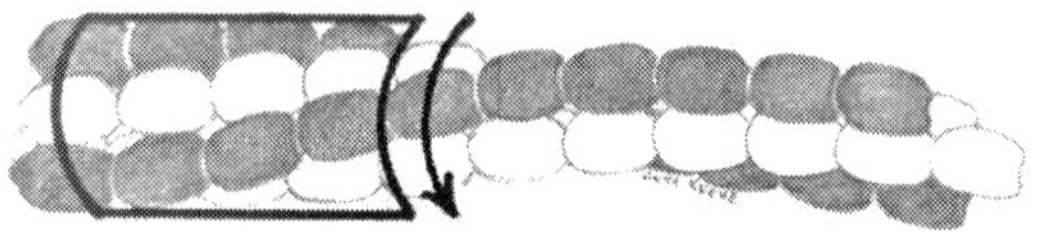

Figure 5.2 The photon spin of the Tor-Chains reflects a differential in Left-Tor and Right-Tor count, inducing an electric field.

The electric field in turn induces and reinforces a magnetic field that extends perpendicular to both the electric field and direction of travel. Together they produce an *electromagnetic field,* as illustrated in Figure 5.3. Because photons are so ubiquitous throughout the universe, the induced EM field is likewise ubiquitous.

The description of the photon's movement also serves to describe the relationship between magnetism and electricity. Scientists have long known that electricity moving through a wire induces a magnetic field. This makes perfect sense if one imagines the spinning electrons moving through the wire headfirst, parallel to their axes of spin. All

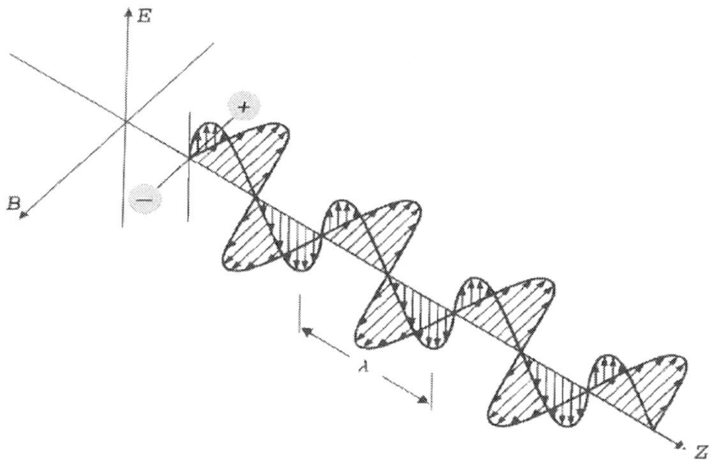

Figure 5.3 The electric field (E) extends perpendicular to the magnetic field (B), and together they create the electromagnetic field.

of those moving, like-spinning electrons in a head to tail arrangement would mimic a Tor-Chain and therefore induce a magnetic field.

Likewise, consider a generator. When a magnet comprised of many particles with aligned spin-axes is rotated, such as an armature, it creates an unequal balance of charged particles, mimicking and inducing an electric field, causing electricity to flow in the adjacent wires. Each of the fields in turn induces the other.

The photon is said to be absorbed/emitted by a charged particle, e.g., a quark or electron. We do not know how this is done. Perhaps the absorption or attachment is due to the charge differential in the charged particle attracting and holding onto the mimicked electromagnetic field of the photon. In any event, the attachment changes the particle's kinetic energy. The photon seems to disappear at the time of absorption, but perhaps it is either attached to or spinning around the particle undetected.

While the universe continues to expand and cool, let's examine an issue we touched upon earlier. One of the consequences of the evolutionary process is the creation of an ocean of various odd particle configurations and a great deal of particle collision debris. In the Tor Model, that debris may be the source of dark matter. To construct our early universe, we must acknowledge the presence and importance of dark matter, but we do not necessarily need to know its origin. But again, for fun let's look at what may have happened to all those exotic particle configurations and their collision waste.

Speculations on the Origin of Dark Matter

Using Tors, Tor-Chains, and Tryk combinations, the evolutionary process could build all kinds and shapes of particles and anti-particles. Stable, long-term creations however, are always limited by the hazards of the environmental frenzy and the competition for survival inherent in natural selection. As is often the case with natural selection, some of the stable creations that do not go on to develop further might continue to exist but remain developmentally stagnant. Accordingly, our medium of thin Penergy will likely contain odd particles and particle fragments that were stable enough to withstand the frenzy, but not flexible enough to become a part of something larger and more complex.

Those odd Tors, Tryks, chains, and particle fragments would have been left behind in the evolutionary development of particles. They remain dark to us because only particles with electric charge can absorb/emit photons of light. Any particle remnant with an attractive or repulsive charge would float around, bumping into other fragments until it combined with the right fragment(s) to neutralize its charge and be unable to absorb/emit photons. Accordingly, you might call these now neutral fragments collectively *dark matter.*

Physicists Paul Davies and John Gribbin expressed a similar idea in their book, *The Matter Myth.* On the origin of dark matter, they said, "Nobody is sure what the invisible stuff is, but the best bet is that it is an unseen residue of exotic subatomic particles left over from the big bang."[18] Some theorize dark matter to be some exotic new particle. It seems more reasonable for dark matter to be normal particle fragments, however, since it is observed that the density of dark matter has the same magnitude as the density of visible matter.[19]

The dark matter fragments would have floated around bumping into each other, gradually expelling their kinetic energy. No longer capable of combining or absorbing more energy, they were left to gather in clumps due to their own gravity. Scientists believe those clumps represent five-times more dark matter than the visible matter in the universe,[20] and that its cumulative gravitational influence played a part in the shaping and distribution of the developing galaxies.[21]

Dark matter fragments possessing little energy and no charge will be difficult to detect. However, if we could isolate a portion of dark matter and spray it with a high energy laser, we might break up the connections, re-establishing its altered configuration from charged particles. This would allow it to once again absorb/emit photons, giving us the opportunity to see it.

Dark matter is the *boneyard* of particle fragments but collectively it has substantial mass and is believed to be adding gravitational influence to our galaxies, which explains the unusual rotational speeds of some of them.[22] There may not be as much dark matter in the universe as scientists calculate, however, as explained in Chapter Six under the heading *Speculations on Dynamic Gravity.*

Another important aspect of particles is the way they are created and decay. Their creation might be related to the presence of dark matter, so while we are being adventurous, let's speculate on particle creation and decay as well.

Speculations on Particle Creation and Decay

Particle Creation

Dark matter particle fragments could remain very useful. After the birth of the universe, once the density of the Penergy drew thin enough, it spun Tor type particles at will. That particle producing density has long since passed. Now it is only when there is a knot of Penergy of sufficient density that it will produce particles. In the Tor Model, the particles it creates depends on the source of the knot of Penergy.

If the knot of Penergy was created in the collision of two high-energy particles, such as two protons traveling at near light speed in a collider, then the knot of Penergy is more likely to create one or more large particles that instantaneously resolve into a flood of lighter particles. That phenomenon is observed daily in particle colliders and is how we discovered the existence of the first and second generation of particles, as well as a host of other unstable particles.

The source of the knot of Penergy could also come from an existing single, high-energy particle, such as a gamma ray. Gamma rays are high energy photons created in explosions, blackhole jets, and other high energy events occurring in the universe.

If the knot of Penergy was created from the natural decay of a gamma-ray photon, the knot will usually spin itself into a particle pair, as described in Chapter Two. But how does that knot of Penergy know what to create, or how to create it? It has the potential to create any particle pair commensurate with its energy content. We can speculate as to how the knot of photon Penergy will respond as it changes into a particle pair.

Perhaps the knot of Penergy would not necessarily spin *fresh* particles. Instead, as the Penergy knot begins to create a pair of left and right-spinning particles, it picks up Tors, Tryks, and fragments from the *boneyard* and uses them either directly, or indirectly as a blueprint, to construct the largest particles possible given the available Penergy.

Scientists have observed a high-energy photon create a matter and anti-matter pair of protons.[23] According to the Standard Model, protons are complex particles made up of two different quarks and a sea of eight different gluons. The anti-proton is just as complex, made up of the same particles with the opposite charge. If two complex matter/anti-matter particles are going to be created out of a knot of Penergy, it is more reasonable that they were rapidly assembled directly, or from a blueprint, of boneyard particle fragments, than having been created anew with such specific complexity simply out of the Penergy alone without such a blueprint. There is no evidence for such a particle creation process, but it is an interesting speculation.

Particle Decay

It appears that particles decay (*decompose*) in two ways, *natural decay* and *forced decay*. A big atom such as plutonium with a large number of nucleons decays naturally by ejecting an alpha particle (*two protons and two neutrons*) from its nucleus, which is a process known as *Alpha Decay*. A down quark decays naturally by ejecting an electron and neutrino, which is known as *Beta Decay*.

Natural decay occurs when an atom or particle becomes unstable. The cause of this particle instability is not well understood. QFT associates the decay with the weak force that encompasses the sudden creation of a very large virtual boson. The Tor Model associates the decay by other means. It could be that in the case of Alpha Decay, the jiggling of a large nucleus occasionally distorts its shape allowing the repulsive forces between protons to overcome the short-ranged nuclear force, enabling the alpha particle to escape.

In the case of Beta Decay, most beta particles are ejected at speeds approaching that of light.[24] Perhaps the absorption of a high energy photon overcomes the binding energy of the electron connection, breaking that connection and allowing the electron/neutrino duo to escape from the down quark.

Forced decay occurs when a particle cannot subsist in the existing density level of the Penergy medium. This occurs in an accelerator or collider when a knot of Penergy creates a very large particle. The large particle immediately decays into smaller particles, and in some cases those smaller particles resolve into even smaller particles, until all particles produced can reside in the existing Penergy density.

Standard Model physicists would argue that the created particles did not come directly from the original particle, but from a virtual boson that mediated the decay. In the case of the decay of a down quark, the boson in question would be a virtual W^- particle. It means that the less than 5 MeV sized down quark first absorbed or suddenly changed into an 80,000 MeV virtual W^- that then immediately changed back into a less than 5 MeV up quark, plus an electron and neutrino. All of that may be possible, but it seems like a laborious way for nature to work.

Such heavy particles have been observed in colliders and physicists have developed the math to support such an inefficient interaction, but relying on heavy, virtual particles to explain particle decay seems dubious. It is also unnecessary if our Sub-A's are in fact composite particles. A more reasonable explanation for the result of the down-quark decay is again that the electron and neutrino were present within the parent down quark before the decay. Admittedly, there is no direct evidence for such a composite configuration, but we just don't have the technology to break down those tiny Sub-A's to that degree, but perhaps someday we will.

As detailed earlier, physicists theorize that after each interaction, which includes a decay, particles are not simply reconfigured but instead all new particles are created.[25] This seems rather extreme, and it makes more sense for nature to handle particle decay like it does atom decay. When an atom decays by spitting out an alpha particle from its nucleus, it simply leaves behind an atom with different atomic numbers and characteristics. The protons and neutrons are reshuffled but otherwise remain unchanged.[26] Particle decay seems more about rearranging existing components than creating all new components from scratch. Nature would work far more efficiently that way.

While we are speculating, let's look at another interesting topic about the universe — special relativity. Einstein gave us the equations showing how space and time are mathematically relative to each other, but the equations do not explain *why* this is so. Let's examine why space and time might be relative to each other. What is the source of that relativity?

Why Space and Time are Relative

The Penergy medium is a wonder to behold. Like the tough fabric it is, it can withstand the contractions, bending, and warping by massive blackholes, and at the same time act as stretched cellophane on which subtle vibrations can be transmitted allowing tiny particles to communicate. It can withstand the comparatively large gravitational waves created by crashing neutron stars, while at the same time hold a single particle's delicate field. It can hold the trillions of individual particles, their vibrational signals, all the photons of radiation and their related electromagnetic fields, and simultaneously allow the tiny signal from my friend's cell phone in Paris, via satellite, to find my cell phone in California, and relay her voice clearly and distinctly. The Penergy medium is truly fantastic and dynamic. It is also closely related to both space and time, and may be responsible for why they are relative to each other.

We can't see or feel space or time, but they remain very real to us. We derive their existence by sensations of objects and the movements of those objects around us. We once thought of space and time as distinct entities but have since learned from Albert Einstein that they are connected. Let's examine how that could be.

A Thought Experiment

Let's do a thought experiment. Imagine a capital T with a railroad track running across its top and a narrow conveyor belt running down its base. A mailman on a passing train wishes to toss a bag of mail onto the conveyor belt. The mail bag resting on the moving train has momentum, which will carry it forward as it falls toward the conveyor belt. Knowing the speed of the train and the drop distance to the conveyer belt, using Newton's laws of motion one can calculate when in advance to let go of the mail bag to hit the conveyer belt.

Now imagine the same setup with the capital T, but this time from the train we are going to emit a photon down the conveyor belt. For the photon to shoot down the center of the conveyor belt, where should the train be when the photon is emitted? From all that I have read, the emission point on the train should be directly adjacent to the center of the conveyor belt. The photon will not travel forward in the direction of the train's motion as the mail bag did. When the photon comes into existence, it immediately steps out above the conveyor belt and shoots away in a *straight path* at light speed, and it will stay on that straight path until influenced by a gravitational field or some form of matter.

This makes sense because whereas the mail bag was *in existence* and subject to the laws of physics regarding the movement of it and the train, the photon *did not exist* in the same sense until it was emitted from the train.[27] Further, the photon was separate from the train the moment it came into existence, so the train had no influence on the photon's movement.

The photon was only subject to the laws of physics regarding the Penergy medium which, except when passing through a clear medium, is the only realm in which a photon ever exists. This goes along with the idea that the speed of light (*the photon*) is always measured the same, whether emitted from a source traveling in the direction of emission or away from the direction of emission. The speed and direction of the source of emission is immaterial to the photon.

Proof that photons travel only in straight lines irrespective of the movement of their source is found in experiments measuring the distance to the moon, where photons are shot to reflective mirrors left on the moon's surface. According to the experimenters, a light beam from earth that strikes the mirrors is reflected back on the exact same path that the beam took to reach the mirrors.[28]

If the light beam shot from a spinning earth is reflected on the same path from the non-spinning moon, then obviously the spin of the earth had no effect on the path of the beam. It is a straight shot in both directions. As physicist

Brian Clegg says, "...light always goes in a straight line. Period. There is no arguing with this."[29]

Another Thought Experiment

The conclusion to draw from the above thought experiment is that a light beam may bend due to gravity, but it will not bend due to the movement of the source of the beam. If this is true, it brings into question an often-used argument for time dilation — the light beam bouncing within a moving spaceship scenario.

The argument is that two mirrors perfectly parallel to each other are set up on the floor and ceiling of a spaceship with a light gun mounted at the upper mirror aimed at the lower mirror. Once the spaceship is under way at high speed the light gun fires light (*photons*) at the bottom mirror. The light hits the bottom mirror and is reflected up, hitting the top mirror and reflected back down, etc.

From inside the spacecraft the photons are simply seen bouncing up and down between the mirrors. For an observer outside the spaceship however, she sees the same light moving in the direction of the spaceship making a series of W's between the mirrors. Given the light is moving in a W, a distance much greater than the direct distance between the mirrors, the photons must be traveling a much longer distance from mirror to mirror. Because light only travels at one speed, the only way for it to travel the longer distance is for *time to slow* within the fast-moving spaceship.[30]

At first impression one may wonder how the photons traveling a lateral distance could be said to be traveling a greater distance necessitating more time than it takes to bounce them up and down between the mirrors. The time it takes for a bullet dropped from five feet to hit the ground is the same for a bullet fired from a five-foot-high level rifle, even though the bullet may travel a thousand yards laterally before striking the ground. Lateral movement does not add any time to falling bullets, why should it for bouncing photons?

That argument aside, let's continue with our thought experiment and say we make the bottom mirror very tiny, perhaps only a few atoms wide. It is still parallel to the top mirror and still capable of reflecting the light gun photons. Now we get the spaceship up to speed and fire the light gun. The light travels toward the bottom mirror, but this time it does not reflect up to the top mirror. It has missed hitting the tiny bottom mirror because the movement of the spaceship has moved the bottom mirror out of the way before the light reached it.

The light when emitted from the light gun was set on a *straight path* downward, and but for the high-speed spaceship moving the bottom mirror out of the way it would have struck the mirror and been reflected. Having missed the bottom mirror, the light does not move in a W or travel a greater distance by being emitted from inside a moving spaceship. If that were the case, light would bend in an arc every time it was emitted from the moving earth, however as discussed above, apparently that is not the case.

Relativity

The above analysis seems correct, but whether valid or not it warrants examining the relativity between space and time a little closer. According to Einstein, the gravitational field of a large mass will cause time to slow. Time is relative to gravity as well as space due to the equivalency between a gravitational field and acceleration. Perhaps time is not relative to space directly, but relative to something equivalent to space that includes gravity. Let's explore what that could be.

Space is the void between stars and galaxies. We call it the vacuum of space, but we now know there is no such vacuum. Space is permeated with dark energy, which we have identified as Penergy. So, when we are saying 'space', we are really referring to Penergy.

For Penergy and time to be connected and relative to each other, it would mean that time is somehow a dynamic of Penergy, perhaps a function of its density. Afterall, time is relative to something changing, and Penergy density seems to control how quickly any change takes place. That may sound crazy, but let's examine the notion and see where it takes us.

If time were a function of Penergy density, then a change in density would mean a change in the passage of time. As we have seen, Penergy was once very dense, but it has been thinning since the universe began. Its density now only changes to a higher density when it is condensed or compressed. Let's examine how those conditions could come about and whether they are related to a change in time. Einstein identified two situations causing time to change: *when near a mass with a strong gravitational field*, and *when flying through space at a very high rate of speed.*

Time Dilation Due to Gravitational Condensing

Einstein's general theory of relativity taught us that the gravitational field of a mass affects both space and time, and that the greater the gravitational field the more space is warped and the slower time passes.

One of the proofs of Einstein's theory is that the GPS satellites orbiting the earth must adjust for the time differential between the earth's surface and their orbital height above the surface, otherwise the system would be thrown off significantly within a short time. This is due to the clocks in the satellites experiencing a much lesser gravitational-field influence than the clocks on the surface of the earth. Time is running slower on the surface than it is in the orbiting satellites twenty thousand kilometers above the earth's surface. [31]

As we have established, a body with mass distorts the geometry of the surrounding spacetime creating an attractive influence. That attraction would likely not only affect other masses, but the surrounding Penergy medium as well. It too would be drawn into the gravitational field, condensing at the surface of the mass. The condensing of the surrounding Penergy would cause it to have a higher density in proportion to the mass's gravitational field. That higher density of Penergy could be what is causing time to slow, not simply the gravitational field produced by the mass.

In the case of a blackhole, the strength of the gravitational field is enormous, having a significant influence on time passage. Theoretically the crew on a spaceship heading into a blackhole would pass through the event horizon without noticing any time change. A person watching the spaceship from a distance, however, would see time in reference to the craft gradually slowing and seemingly come to a stop as the spacecraft reaches the event horizon.[32] That notion has obviously not been validated, but the point is that the stronger the gravitational field, the slower time passes.

Penergy does not have mass as we define it. It's fluid-like consistency makes it respond differently to a blackhole's gravitational field than does mass. While mass is drawn in and consumed by the blackhole, Penergy is drawn close but is largely pushed away at the event horizon by the spin of the blackhole. If the blackhole is rapidly spinning, as most do, the Penergy is pulled in tightly to the blackhole, but is held at bay. Evidence for this is in the fact that if fast spinning blackholes eat Penergy like they do objects with mass, the first blackholes would have likely consumed all the remaining Penergy in the universe and we would not be here.

The fast-spinning blackholes drag the surrounding Penergy with it, performing the *frame dragging* described earlier. The surrounding Penergy builds in density as it is pulled close, causing time to slow as seen by the observer watching the rocket approaching the event horizon.

The much higher Penergy density at the event horizon likely also causes particle pairs to be created consistent with the strength of the density. Those particle pairs include the first and second generation of particles — the heavier quarks and leptons. Some to those particle pairs would be drawn into the blackhole, and some would escape, being slung out into space. This could be why the earth is constantly bombarded by cosmic rays — high energy particles and anti-particles from space having an indeterminant origin.

Time Dilation Due to High-Speed Compression

The other circumstance in which time is dilated according to Einstein is when an object such as a rocket is traveling rapidly through space. The closer the rocket is traveling to the speed of light; the slower time passes for those within the rocket. There could be a couple of explanations for how speed is related to Penergy density and the passage of time.

According to Einstein, a mass such as a rocket traveling through space [*the Penergy medium*] would grow in mass size the faster it traveled.[33] The high-speed mass would cause the surrounding Penergy to compress. The compression of the Penergy would raise its density in proportion to the mass's speed and growth in mass size, and consequently slow time for those inside the rocket.

Another explanation, through possibly related, comes from the idea Penergy spins itself into particles. That spinning mass then affects, and has a relationship with, the surrounding Penergy. If photons are added to the particle to move the particle/mass faster through space, the particle's spin rate goes up, changing the relationship between the particle and surrounding Penergy.

The higher rate of spin pulls on the Penergy, causing it to get a bit denser around the particle. The more photon energy that is added to speed up the particle/mass, the faster the spin rate and the denser the Penergy gets, and consequently the slower time passes for that particle/mass. In this way particle speed is related to a change in the passage of time, but it is actually the added cumulative spin rate that is changing the Penergy density, that is in turn causing the change in the time passage.

There is no evidence for either of these time dilation theories being related to Penergy density, but they do represent a possible nexus between time, gravity, and space. If proven valid someday, it would mean that Penergy truly is *spacetime*.

Entropy — Cosmic Recycling

Thermodynamics is the study of heat and energy. The reader is probably familiar with the first law of thermodynamics: energy cannot be created or destroyed. The second law of thermodynamics encompasses the idea that things and systems are usually wearing down, becoming less energetic and more disordered and random. The term to describe this phenomenon is *entropy*. There are many interpretations of entropy and what follows is only one of many perspectives on the subject.

The Combination and Growth Imperative (CAGI) progresses in evolutionary stages to produce Levels of matter that are put together (*combined*) from the elements in the Level that preceded it. In this way, matter is constructed from building blocks that are constantly growing in size and complexity.

A growth in complexity necessitates a variety of building blocks. Those building blocks must be readily available and capable of combining with others to test which combinations are best suited for the environment in which they reside. The best building blocks are of course those that have already endured the tests of a punishing, hazardous, or frenzied environment. The best are those that have been recycled; fragments tossed into the boneyard to be reused when needed.

Entropy is the universe's way of breaking down complex units into smaller, less energetic building blocks, making them available for reuse when needed. Even the highly complex levels created by the CAGI process are subject to entropy, and for good reason. If nothing broke down at each level of complexity, we would soon run out of choice building and energy blocks. The environment is constantly but subtly changing, requiring constant entropy to produce building blocks with the latest core, survival attributes. For complexity to grow effectively, death and decay must be a part of the process. No level or their building blocks are permanent. They are used by the CAGI process temporarily to further its progress and then recycled. It's just how evolution and the universe must work.

* * *

In this chapter we looked at different particle fields and concluded that the Penergy medium would make a better backdrop for particle fields than the mathematically based Quantum Field Theory that prescribes separate, ubiquitous, undetectable fields for each type of particle. We then speculated on the origin of dark matter, particle creation, and particle decay. We concluded the chapter with a look at why space and time might be relative through a change in Penergy density.

We will now move on with our construction of the early universe, addressing the creation of the fourth level of building blocks — atoms.

Chapter Six

Level Four Building Blocks

As the universe expands and cools, the existing energy level comes down to the point where subatomic particles can finally begin to combine, but the frenzy is still pretty wild. There is still a good deal of evolution to take place before atoms form. To be clear on how subatomic particles combine let's review the forces that affect how particles interact. We discussed particle fields in Chapter Five as being unique disturbances in the Penergy medium that surrounds the particle. We will now examine *force*, which is the effect those fields have on other particles.

The Forces

As mentioned earlier, for particles to combine they must interact through one of the known forces. In the Standard Model, the four forces are theoretically mediated by virtual bosons: photons, gravitons, gluons, and W and Z particles. Particle interaction is governed by certain rules that describe the four forces.

According to Mathematician Milo Beckman, the exact rules for particle interaction in the Standard Model are absurd. They amount to a complex, systematic process, but not a neat and simple one.[1] Physicist Michio Kaku tells us the Standard Model was created by splicing together by hand the theories that described the various forces, so the resulting theory is a patchwork. He reports that one physicist compared it to taping a platypus, an aardvark, and a whale together and declaring it to be nature's most elegant creature.[2]

Make no mistake, the Standard Model's equations describing particle interactions produce accurate predictions. Dr. Kaku's imagery, however, is a good reminder that the Standard Model often makes the math work, while creating a picture from its interpretation that is questionable. Let's examine each of these forces to assure our theory of particle combinations is on sound footing.

Electromagnetic Force

The electromagnetic force is a combination of the electric and magnetic forces. The electric force involves the attractive and repulsive effect between particles possessing *electric charge.* We have already examined the electric force in Chapter Four and concluded that its source is more likely through the fields of mirror-image Tor particles rather than the constant exchange of virtual photons. Photon exchange may work mathematically, but because it requires the presence of undetectable, virtual particles conveniently popping in and out of existence with no clear way to exert an actual attractive/repulsive force, that theory is dubious.

As discussed in Chapter Four, the attractive/repulsive forces of *charge* are the same forces found in the attractive/repulsive aspects of matter/anti-matter and the magnetic force. The magnetic force only involves like-spinning Tors, and the attraction is only affected when Tor particle fields are aligned nose to tail but are otherwise repulsive. Since

Maxwell brought the magnetic and electric forces together under a single set of equations, the two forces have been thought to be two sides of the same coin. In the Tor Model, that common coin from which both forces originate is in the fields of the Tor particle, most likely due to the dynamics of the Tor's spin.

The electromagnetic force is responsible for the attraction between electrons and protons, a combination necessary to create atoms, for atoms to create molecules, and for molecules to create life. The electromagnetic force is also what we see in action all around us daily. The repulsive aspects of the electromagnetic force are what stops us from falling through the floor. The molecules in the floor are holding on tightly with the electromagnetic bonds that repel our shoes, holding us above the floor.[3] Obviously, the electromagnetic force is very important in our daily lives.

The Force (*F*) between two charged particles falls off inversely to the square of the distance between the two charges (*q*). The equation expressing this notion is, F=Kqq/ **r2**; (*K being merely a constant*). It's the $/r^2$ that represents the inverse square part of the equation, meaning at twice the distance, $/2^2$, the force is $1/4^{th}$ of what it was, and at 3 times the distance, $/3^2$, the force is $1/9^{th}$ what it was, etc. This point is important to remember as we turn to the discussion of the gravitational force.

Gravitational Force

According to the Standard Model, *Gravitons* are the particles mediating the gravitational force. That means that all the particles in your body and every other earthbound body are constantly exchanging gravitons with the earth. It would mean the earth is constantly exchanging gravitons with the moon and sun. It would mean the sun and every other star in the Milky Way is constantly exchanging gravitons with the SMBH at the heart of our galaxy. It's seemingly the only way all those gravitational forces could exist if gravity worked by exchanging gravitons. Wow, that is a lot of long-distance particle-exchanges constantly going on!

The graviton-exchange scenario is difficult to believe. The existence of gravitons has never been detected, and besides Einstein gave us a much more plausible explanation for gravity. Einstein showed us that gravity arises out of mass distorting the fluid-like nature of space, and that the distortion affects the movements of nearby matter relative to the size of the respective masses. Einstein's General Relativity tells us how gravity works, but not what it *is*. What is causing space to be distorted? We will speculate on that question in the next section.

Interestingly, the strength of the gravitational force falls off at exactly the same rate as the electromagnetic force: inversely to the square of the distance between two masses (*m*).[4] The equation expressing this notion is F=Gmm/**r2**; (*G being merely a constant*). The similarity in force dissipation between these two forces must be more than a coincidence. Two distinctly different forces that lose strength at the exact same rate suggests they have some underlying feature in common. Let's explore what that feature may be. To build our universe we need to understand how gravity works but not necessarily its source, but again for fun let's speculate.

Speculation on the Source of Gravity

From Chapter One we learned that the only two distinct stable things in the universe are blackholes and Tor particles, both of which can possess mass. We know gravity is an attractive force due to the warping of the fabric of space (*the Penergy medium*) in proportion to that mass. But how does shear mass distort space and exert a gravitational field? Perhaps it is not the mass itself that is warping space, but something else that exists *in proportion* to the mass.

We know that massive, spinning objects such as blackholes distort space.[5] This observation could be important. It might be the *spin* of the object that is pulling on and distorting space, not simply the object's mass. That may work for blackholes that are known to possess significant mass and spin, but how would that work for tiny particles that are only part of a massive object, especially one that may be spinning relatively slowly such as a planet?

Mass is simply an accumulation of particles, and those particles are all made up of combinations of Tors. The only way any mass has contact with the Penergy medium is through the existence of Tors, and the contact point is the surface of those spinning Tors. Tors spin on their own axis and presumably at a very high rate. Tor spin therefore might be capable of tugging on the Penergy medium to distort it causing a gravitational field. Perhaps gravity is no more than the cumulative effect of all the Tor spins in a mass tugging on the Penergy medium.

There are two types of mass. Mass that is measured by its resistance to movement is called *inertial mass*. Mass measured by its gravitational field strength is called *gravitational mass*. These two types of masses are based on different concepts, but they always measure exactly the same. It has been a mystery as to why this should be so.[6]

As noted in Chapter Three, the source of inertial mass may be due to the gyroscopic effect created by the very-fast spinning Tors. We now see that the source of gravitational mass could also be due to the spinning of Tors. The Tor particle could very well be the nexus between mass and gravity and be the reason inertial mass and gravitational mass have the same value.

Recall that the source of electric force is ultimately due to the spin dynamics of Tors. Now we are saying the nexus between mass and gravity could also be due to Tor spin.

This could also be the reason the dissipation of the field strength of the electric and gravitational fields (*inversely proportional to the distance between charges/masses*) are identical. Both are derived ultimately from the same source, Tor spin.

The attractive force of electric charge occurs *directly* between Left-Tors and Right-Tors, each sensing each other's field. However, the attractive force of gravity occurs between the spin of Tors and their cumulative effect of warping the surrounding Penergy medium, which *in turn* affects other particles. This *indirect* effect of gravity on other particles could be the reason why it is much weaker than the other forces.

As mentioned in Chapter Five, some of the added gravitational influence effecting galaxy rotation could be coming from the supermassive blackholes at the heart of the galaxies. In the Tor Model, their gravitational strength can be greater than their mass if the denser Penergy fuel inside of them drives their spin rate up disproportionately creating their own *dynamic gravity*.

Speculations on Dynamic Gravity

The strength of a blackhole's gravitational field is believed to be proportional to its mass. A blackhole's mass is calculated based on its effect on other objects such as surrounding stars and gases. In the Tor Model, the strength of a blackhole's mass and gravitational field is proportional to the density of its Penergy, which in turn has a relationship to its spin rate.

According to Einstein, mass couples to spacetime,[7] and as previously observed massive spinning objects such as blackholes are known to distort space. It is quite feasible that it is the spinning of the blackhole that tugs/distorts the Penergy medium and has a significant influence on, if not creating entirely, the gravitational field, as speculated in the previous section.

Prediction: Someday an enterprising young physicist will figure out a way to measure the gravitational field surrounding a SMBH and discover that the strongest area of its gravitational field is at its equator, where it is spinning the fastest. This will confirm that it is not simply the mass of the SMBH that is creating the gravitational field, but the spinning of the SMBH.

As mentioned in Chapter One, it makes sense that matter inside a blackhole goes through transitional phases as it becomes denser. The blackhole crushes matter back into Penergy, the Penergy into denser and denser Penergy, and finally into its NIB (*Near Infinite But* not quite) state. Stephen Hawking told us that when something falls into a blackhole it would be trapped and would collapse into some unknown state of very high density.[8] This is consistent with the Tor Model's vision of the inside of a blackhole.

In the Tor Model, gravity can only crush matter down to a certain level, to the point when it is no longer an independently spinning entity but becomes a glob of fluid-like Penergy. Beyond that point the further condensing is not due to gravity per se, but due to the dynamics of the fast-spinning Penergy itself.

The heart of the blackhole is a long distance from, and no longer influenced by, the gravitational contraction of the surrounding thin Penergy medium. Inside the blackhole the fate of the Penergy being condensed is now determined by the dynamics inherent in the Penergy condensing process, taking it into denser and denser phases. This state is beyond the realm of our current version of gravity described by general relativity, so it will require further investigation and perhaps a new mathematics before it is understood.

The radius to which a mass must be compressed down to in order to form a blackhole is called the object's Schwarzschild radius, named after Karl Schwarzschild, who was first to find a solution to Einstein's equations describing a spherical blackhole.[9] The larger the mass of the blackhole, the larger the Schwarzschild radius.[10] In the Tor Model, this proportionality eventually changes.

The effect of the slow building NIB-state is called *dynamic gravity.* It is supplemental to that of normal gravity created by the contact between the spinning blackhole and the Penergy medium. Dynamic gravity may not manifest itself until the blackhole is quite large and has processed a tremendous amount of matter and Penergy. The tell-tale sign of a NIB-state would be when the blackhole's spin rate goes up and its event horizon begins to shrink. Some SMBHs may be in the initial phase of that state already, as spin rates have been measured to be fifty to ninety-five percent of light speed.[11]

The spin-rate/circumference relationship is subject to the same dynamics an ice skater employs to spin faster by pulling in her arms, shrinking the radius of her mass. In the blackhole's case, it is the faster spin rate that is causing the radius of the mass to shrink. Theoretically, if we fed all the matter and Penergy in the universe into a single, giant blackhole, its spin rate would continue to rise, its circumference would continue to shrink, and eventually it would return to the speck of NIB-density Penergy from which our universe began.

Scientists have put forth theories that support stronger gravity and less reliance on dark matter to explain unusual galaxy rotation. Israeli Physicist Mordehai Milgrom proposed simple changes to Newton's laws that when applied to galaxies turned out to match their rotation speeds quite well. His theory became known as MOND for Modified Newtonian Dynamics. MOND theory is out there but is not yet well accepted. It did, however, show that the mass of a galaxy is related to its rotational velocity.[12] Perhaps the final theory affecting galaxy rotation will be something in between MOND, dark matter, and Dynamic Gravity. Keep in mind there is no evidence yet of dynamic gravity; it is only speculation. But someday an enterprising young physicist may develop the mathematics to support such a theory.

Strong Force

According to the Standard Model, the three quarks of a nucleon are held together by the constant absorption/emission of virtual gluons. This force is unique in that no matter how much energy is put into the effort; the three quarks cannot be separated. It is as if they are held together by strong rubber bands.

Quarks and gluons are theorized to possess a special kind of charge called Color Charge, which controls how quarks exchange gluons. A color charge has never been detected and is derived solely from a brilliant scheme developed by physicist Murray Gell-Mann in the 1970's that explains very accurately how this force might work. The Strong Force operates at very short distances and only inside the nucleus.

The theory and math supporting it are quite complex, there being three different color charges and eight different gluons involved. A good summary of quarks and gluons and how they were theorized can be found in Timothy Paul Smith's, *Hidden Worlds — Hunting for Quarks in Ordinary Matter*[13].

In the Tor Model we established that one of the ways particles combine is by sharing a part of themselves, like atoms sharing an electron to create molecules. One could argue that quarks are sharing a part of themselves, a gluon, but it's not quite the same. When atoms share electrons they create a momentary charge imbalance, and it is the charge imbalance that is actually holding the atoms together, not the mere sharing of the electron. In the case of the gluon exchange, because gluons are theoretically electrically neutral, sharing a gluon would not create an electric charge imbalance.

The Standard Model overcomes this by theorizing the Color Charge and the need for a composite particle, e.g., a proton, to be 'color neutral'. That means a composite particle having three quarks must have one and only one of each of three colors per quark. A composite particle comprised of two quarks must have a color and an anti-color in the two quarks.[14] The source of color charge is unknown.

It remains difficult to believe there is a force that holds two or more particles together by exchanging virtual particles. There is also the argument again that every time there is an absorption or emission, all particles involved are destroyed and all new particles produced, creating the height of inefficiency.

Scientists now believe a nucleon contains far more matter particles than just the three quarks defining its electric charge value. The kind of particles involved is unknown. This makes sense because the mass of a nucleon is over one hundred times the mass of the three defining quarks combined, so there must be more to the nucleon than presently theorized.

The Tor Model offers no explanation for how the three defining quarks are held together. The only force it recognizes acting like a rubber band would be the very-strong magnetic-force working at the Tor level. As observed earlier, the creation of Tor-Chains is due to the powerful coupling-capacity of the aligned fields of Tors spinning on their own axes. Such a force might act like a rubber band if one tried to pull apart even a lengthy Tor-Chain. This would only come into play if the quarks were joined by Tor-Chains, but experiments suggest the quarks might be three independent particles.

Otherwise, the Tor Model can only speculate that their connection is probably due to the sharing of particles between all the internal parts of the nucleon. A complete theory may be possible once the internal parts of the nucleon have been better defined.

Weak Force

A stand-alone neutron is unstable and will decay into a proton, electron, and neutrino within a few minutes. By the turn of the twentieth century scientists were familiar with the magnetic, electric, and gravitational forces, so to explain how a neutron could decay into a proton scientists had to define one more force.[15] A new particle had to be introduced to explain this force.[16]

Large particles were predicted and ultimately detected in the Large Hadron Collider and were dubbed the W and Z particles. They have not been observed directly but are believed to have been identified by reading particle collision debris.[17] According to the Standard Model these W and Z virtual particles are the force carriers of the Weak Force that is responsible for the decay of the neutron and all other particle interactions inside the nucleus.

The decay of a neutron into a proton is actually accomplished by a down quark inside the neutron decaying into an up quark. In the Standard Model quarks are elementary particles, yet the down quark is known to decay into an up quark, electron, and neutrino. Since the down quark is an elementary particle, the only way for it to decay into smaller particles would be for it to first change into a W^- particle that would then decay into the smaller particles.

As mentioned in Chapter Five, the W and Z particles are comparatively large, being more than 80,000 MeV, compared to the light-weight quark at less than 5 MeV. The W and Z particles exist only briefly before decaying. They are only seen in the Large Hadron Collider under very high energy conditions, yet the Standard Model has them coming into existence every time there is a decay/creation or absorption/emission within the nucleus in moderate energy conditions.

Physicists have the math to support the notion these very massive particles come into existence by momentarily *borrowing* the requisite energy from the *vacuum of space*. As noted in our discussion on fields in Chapter Five, apparently the available energy for all those borrowings is not actually present in the vacuum of space.

These massive particles may momentarily exist in colliders, but they might simply be only first or second-generation creations that could not continue to exist in the thinning Penergy density. They may have gone the way of all other first and second-generation particles, making their relationship to today's particles questionable.

A more reasonable approach to down quark decay as explained earlier is for the parent particle to be comprised of the residual particles. Since composite quarks of the Tor Model do not require a special particle to decay, and the Tor Model's belief that down quark decay may be triggered by the absorption of a high energy photon, there appears to be no value in further guesstimating the W and Z particle's makeup or purpose.

Let's continue with our construction project by putting together the next Level of matter. For atoms to evolve, first there had to evolve a nucleus of protons and neutrons. Let's examine how those nuclei likely came about according to both the Standard Model and the Tor Model.

Evolution of a Nucleus

According to the Standard Model

According to the Standard Model, the following points theorize protons and neutrons. They were created within the first second of the Big Bang and came together to create nuclei during a period/process called Primordial Nucleosynthesis. Unconfined neutrons decay in less than fifteen minutes. In the initial high energy conditions following the Big Bang protons and neutrons could change identities converting back and forth into each other, but these conditions were short lived, coming to an end as the universe expanded.

Once the conversion stopped, the universe had only fifteen minutes to create the nuclei, otherwise it would have run out of neutrons making it unable to create atoms larger than hydrogen.[18] The protons and neutrons immediately began combining to create nuclei with different configurations, primarily hydrogen, helium, and some of their isotopes. All the existing neutrons were taken up into nuclei within a few minutes of the Big Bang.[19]

This sequence is part of the Hot Big Bang theory. That theory may be feasible mathematically, but it makes more sense for complex particles to have evolved from something smaller, and for each aspect of evolutionary development to follow from a reasonable cause without the need of anything magically popping into existence. Let's continue with the story of how nucleons and atoms evolved, but from a more natural, evolutionary sequence.

According to the Tor Model

Both models agree on the basic configuration of the proton being comprised of two up quarks and one down quark. Those quarks have charges of +2/3[rds], +2/3[rds], and -1/3rd respectively, giving the proton a +1 net charge that perfectly offsets the -1 charge of the electron. This allows for the eventual creation of a neutrally charged hydrogen atom composed of one proton encircled by one electron.

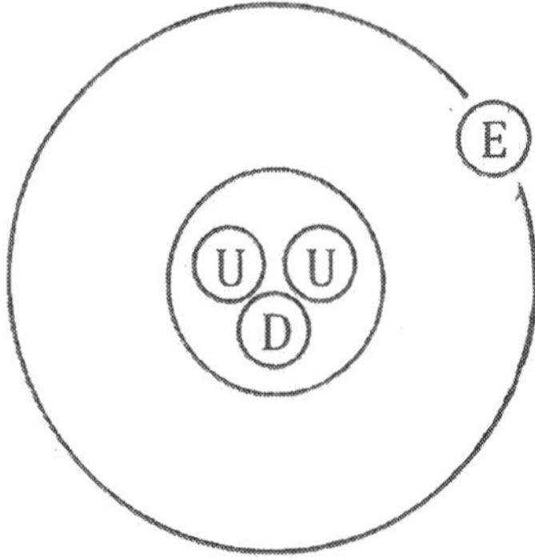

Figure 6.1 A hydrogen atom is comprised of a single **E**lectron encircling a single Proton, which is comprised of two **U**p Quarks and one **D**own Quark.

The reader may recall from high school science class that each element on the periodic table is distinguished by the number of protons in the nucleus, and that two or more protons, because they all carry a positive *charge,* will naturally repulse each other. Quarks being held together by sharing gluons spill over to protons sharing gluons. Sharing gluons helps hold the protons together by overcoming their natural repulsive force. Apparently overcoming that repulsive force between protons was insufficient for the gluon exchanges alone, necessitating the involvement of another particle — the neutron.

A neutron is comprised of two down quarks and one up quark, giving it a neutral (*-1/3, -1/3, +2/3*) net charge. Up quarks and down quarks are similarly configured with the down quark having a little more mass that stems from possessing the additional components of at minimum an electron and neutrino. Should the down quark decay by giving up that electron and neutrino it becomes an up quark. The change in quark count turns a neutron into a proton. The opposite can also happen when an up quark gives up a positron and neutrino turning a proton into a neutron.[20]

Protons and neutrons can exchange a part of themselves, a two-quark meson called a pion. Pion sharing allows the protons and neutrons to rapidly exchange identities, which also helps to overcome the natural repulsive forces between the protons.[21] Neutrons apparently would have to evolve to become a neutralizing agent in the nucleus of atoms larger than the hydrogen atom. Presumably this was necessary to give the nucleus more product from which to make pion and gluon exchanges.

Given that neutrons naturally decay into protons in a matter of minutes, it makes more sense for the hydrogen atom to have already evolved when neutrons evolved, which allowed neutrons to readily join a nucleus before decaying. Once inside a nucleus, neutrons become quite stable. It makes no sense for neutrons to have evolved before the hydrogen atom because without the need for the neutron in atoms larger than hydrogen, the neutron would have no reason to evolve. There would have been no purpose driving its evolution.

Evolution of the Atom

Once the four stable subatomic particles (*photon, electron, neutrino, and quark*) evolved, the evolution of the hydrogen atom probably went smoothly but not necessarily quickly. After the photon evolved from simple Tor chains, the next stable particles were probably made up of Tryk chains. Evidently the Tryk combination creating the electron was quite stable and naturally selected early.

To be instrumental in the next level of matter, its negative -1 charge needed an offsetting positive +1 charge particle. The anti-electron, the *positron*, has a positive +1 charge value, but because it is a mirror image of the electron they annihilate, making it unsuitable for the off-setting task. Ultimately it was the evolution of various quark combinations with 1/3rd and 2/3rds charge values that the proton with a positive +1 charge value evolved.

From there, neutrons would evolve having nearly an identical configuration as the proton. Neutrons gave product and stability to a nucleus allowing heavier atoms to evolve. The high energy gluon/pion exchange forces could bring protons and neutrons together to create the larger nuclei of deuterium, helium, lithium, and beryllium. No other particle combinations were necessary to create the atomic level of matter.

The Sequence of Particle Evolution

All particles are created from Tors.
Individual Tors → Tor Chains → Photons (and possibly Neutrinos)
Individual Tors + Tor Chains → Tryks & Tryk Chains
Tryk Chains → Electrons & Quarks (with +/ — 1/3rd & 2/3rds charge)
Variously Charged Quarks → Protons
Protons & Electrons → Hydrogen Atoms
Proton + Electron + Neutrino → Neutron
Protons & Neutrons & Electrons → Helium and larger Atoms

Because the universe was relatively large before particles were created, the mass-energy and temperature scale of the early universe was high, but perhaps not nearly as high as prescribed by the Standard Model. The combining and building in complexity of all these particles likely transpired using naturally occurring conditions and forces, without the need for ultra-high temperature, though at times the universe would have been very hot.

Once atoms began to evolve, the Standard Model and Tor Model's picture of the universe is quite consistent. The next step in the universe's construction commensurate with the creation of atoms was the liberation of photons creating the Cosmic Microwave Background (CMB).

The CMB

According to the Standard Model, when the early universe eventually cooled to 3000K and allowed electrons, protons, and neutrons to combine to form atoms, it freed photons that had previously been bound up in those particles.[22] This sudden liberation of photons universe-wide sent them soaring throughout the expanding medium. Most of those photons are still flying along untouched.[23]

Since the universe has expanded since that time, the wavelengths of that thermal radiation were stretched by the expansion causing the radiation energy after 13.8 billion years to cool down to just a few degrees above zero.[24] The remnants of that event can be seen today by the presence of a background of very cool 2.75K microwave photons that uniformly exist in every corner of the universe. Those remnants are called the *Cosmic Microwave Background* or the CMB. Their presence is one of the cornerstones of the Big Bang theory and one of the yardsticks for measuring both the age of the universe and the rate of its expansion.

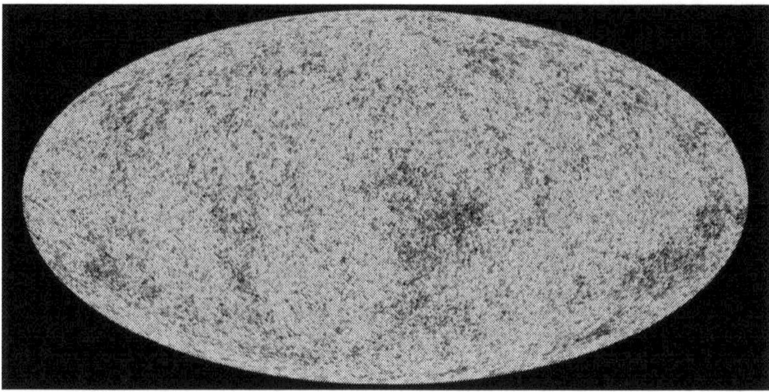

Figure 6.2 The Cosmic Microwave Background; provided by coolcosmos.ipac.caltech.edu. The light and dark areas show a very slight temperature difference due to fluctuations in the conditions of the primal universe.

The existence of the CMB is consistent with the Tor Model's version of the early universe. Both models prescribe an era of particle creation followed by the atomic creation and the liberation of photons. The Standard Model places the time period for creation of the CMB at 380,000 years after the Big Bang. Although the Tor Model's timing may work out to be different, the validity of the creation of the CMB appears reasonable in both models.

In their book *Cosmology for the Curious,* Delia Perlov and Alex Vilenkin tell us that one of the problems with the Standard Model is that to account for the existence of clusters and super clusters of galaxies, scientists must postulate the existence of small density fluctuations which gradually evolve into those structures. But they have no answer for the origin of those small density fluctuations.[25]

In the Tor Model those fluctuations could be caused by the gradual disappearance of the intermediate Penergy swirls that left large holes in the galactic structure. Those swirls would have been disappearing at the same time particles were being created and perhaps even up until the time atoms were created and photons liberated. Those gradual changes in the galactic structure may have created the conditions that we interpret today as fluctuations in the primordial universe. Of course, in the Tor Model such fluctuations are unnecessary because the creating of clusters and superclusters of galaxies is accounted for in the Cascade Effect described in Chapter One.

Possible Alternative to the CMB Redshift

The CMB photons are considerably redshifted, meaning they have lost much of their energy. The Standard Model accounts for the redshift with the notion the photon wavelengths were lengthened due to the photons laboring against the expansion of the universe during the billions of years those particles have been flying along. That may well be the correct answer, but there may be an alternative explanation.

The universe was much smaller when the SMBHs were created and the CMB photons were liberated. It is possible that though all of the SMBHs may not have been fully developed, they nonetheless had sufficient overall gravitational influence to require the photons to lose energy climbing out of the surrounding gravitational fields for much of the early part of their journey.

Mathematics and Uncertainty

Since mathematics plays such a big part in the theoretical developments and description of the workings of the universe, we should get a perspective on its evolution as well. The creation of particles not only created our cosmic dimensions but other attributes of the universe that have been observed, measured, and reduced to a mathematical language.

Like any language, math has rules, and those evolving rules are often created to fulfill a need, sometimes stretching the logic of the rules. New math to fulfill a need is wonderful in its usefulness, but it creates a bias in its language in describing a desired outcome. Consequently, though math has played a fantastic role in our understanding of the universe, it cannot always be relied upon to describe the exact picture of nature, though occasionally it gives us glimpses of aspects of nature we would not have otherwise recognized or even entertained.

The history of math might be broken down into four mathematical realms, each of which were created to describe new discoveries and theories. The first is the *classic realm* wherein physics was first born. This is sometimes referred to as Newtonian math, as Sir Isaac Newton is credited with creating/organizing the laws describing the motion of things, the way gravity plays a part in their movement, and the math describing those movements. The laws and math relating to Newton's work is also referred to as classical physics.

Classical physics and math worked well until the beginning of the twentieth century when Einstein showed us such things like matter and energy are one and the same ($E=MC^2$), light can be both a wave and a particle (*photons*), time and space are relative (*Special Relativity*), and that gravity and acceleration are equivalent (*General Relativity*).

This required new concepts in math initiating the *General Relativity* realm, which is used to describe the larger and faster moving aspects of the universe. At this point our classical picture of reality begins to disintegrate. Space and time, once thought to be independent aspects of the universe, were now joined together in a combined concept called *spacetime.*

Einstein's work laid the foundation for further discoveries in the subatomic arena where scientists were exploring matter at smaller and smaller scales, until they could no longer measure the location and momentum of a particle simultaneously. Neither classical nor Relativity math worked at this level. Realizing experimental results could only be predicted as a *probability* of an outcome, a whole new mathematical realm named Quantum Mechanics (QM) was developed.

Though QM gave only probabilities, it nonetheless proved to be quite accurate in describing the very small (*quantum*) aspects of the universe. Its development caused our picture of reality to disintegrate even further. Realizing particles were waves without a precise location, coupled with an inability to measure their location and momentum simultaneously, meant that knowing a particle's momentum also meant that its location could be *anywhere*. Our picture of reality had definitely become weird.

Science has delved deeper into matter since then but not so deep as to require yet another math. QM is still the reigning math realm, but it has been given more depth over the years. Scientists recognize the potential for going deeper into matter at levels a bit closer to 10^{-35} meters, which is referred to as the Plank realm, named after Physicist Max Plank. This level may produce new attributes to matter that none of the first three math realms can adequately describe, so if science ever gets to the Plank level of time, space, and matter, it may necessitate a whole new body of math that might be called *Plank math.*

Much of the Standard Model's mathematics regarding particles and forces gives very accurate results. Except for the consequences of starting the universe with a singularity and positing particles and forces inexplicably popped into existence, the Tor Model should easily adopt the Standard Model's mathematics regarding the measurements and relationships between particles and forces. Again, it is not the core mathematics the Tor Model takes issue with, but the interpretation of much of the mathematics the Tor Model finds questionable.

The Tor Model may not necessitate a new math, but once its perspective on the evolution of the universe is evaluated and tested, physicists may have to create new terms and values for Penergy, Penergy density, and the multiple aspects of a particle's field. We may need new mathematics to describe the dynamic interior of a blackhole as well. Einstein's equations are fine for the concept of mass as we know it to be equivalent to our common concept of energy, and General Relativity might describe a blackhole to a limited extent, but once the Penergy is condensed into deeper and deeper densities approaching the NIB state, it could be well outside the scope of General Relativity, requiring a new image and new mathematics.

We have built an early universe through observations and deductions that blend smoothly into the universe believed to have existed 380,000 years after the Big Bang. What makes this reconstruction of the early universe compelling is that we did it without anything exploding, popping into existence, or inflating. There was quite a bit involved, but every effect we observed was a consequence of a reasonable cause. That is how evolution works, from a rational sequence of causes and effects.

* * *

In this chapter we looked at the four forces prescribed by the Standard Model and again concluded that the basic attractive/repulsive force inherent in matter and anti-matter, electric charge, and the magnetic forces are all, arising out of the relationship between the fields of the Tor particle. We then speculated on *dynamic gravity* and the importance of Penergy density in measuring and understanding a blackhole.

We then outlined the evolution of the neutron and brought together the subatomic particles to create the atom and followed that with the creation of the Cosmic Microwave Background, which is the point where the Tor Model and Standard Model come together.

With our construction project complete, we need now only look at what that construction process has been about. What is it is telling us about the universe? What is it telling us about *reality*?

Chapter Seven

The Road to Reality

We have completed our construction of a sound theoretical foundation and the initial supporting levels to our early universe. The Model we created is, again, not meant to represent exactly how the early universe evolved but only how it could have evolved. We may never be able to prove the existence of Penergy, pure energy blackholes, or early elementary particles, but their existence answers many questions and seems quite reasonable given our observations.

We shall now take a moment to evaluate whether we are appropriately interpreting this new vision on the early universe. Stepping back and taking in the breadth of our project we see billions of spinning, pure-energy, supermassive blackholes that were initially surrounded by very tiny, early elementary particles. Those particles have combined and grown in Levels, building in complexity. Revealed at each Level were previously unseen *emergent qualities* that have included wondrous things such as fields, forces, the ability to absorb/emit photons, and the variety of atomic elements that will be used to build our universe from here.

That is what we see on the surface, but is there more going on beneath the surface? Can the presence, behavior, and interaction between all these components give us a picture of a deeper reality? Can our Standard Models and the equations of general relativity and quantum theory be the path to discovering a deeper reality? According to physicist Lee Smolin, physics should be more than a set of formulas that predict what we will observe in an experiment; it should give us a picture of what reality *is*.[1] The discoveries and theories of particle physics and cosmology have given us road signs pointing to a subsurface reality. Let's head down that road to reality and see where it takes us.

Finding Reality

Let's start with a caveat: we are limiting our definition of reality to what is happening in the universe independent of a life-form or conscious observer. This allows us to sidestep all the philosophical issues regarding the fallacies of our perceptions, limits to our sense organs, etc. The reality we seek is the one that is happening independent of us.

For physicists and cosmologists, reality is ultimately expressed in equations. So far, matter, energy, fields, forces, their movements, and relationships have lent themselves to being described quite accurately in mathematical terms. That success says a great deal about human curiosity and ingenuity. We seek to understand our world. We peer into the depths of space and the Levels of matter in hopes of learning the laws of nature that will explain who we are, where we have been, and where we are going. We are a part of a mysterious nature, but one that can apparently be understood. There is much we know about it, but much we don't know. Given the success of our understanding so far and the success of our describing it mathematically, we believe there must be a description of the single reality that encompasses all there is.

The Standard Model of Particle Physics and the Standard Model of Cosmology are touted as being two of the greatest accomplishments of the twentieth century. Given what they encompass and what they have accomplished, they are indeed monuments to the human endeavor. The math of these Standard Models apparently isn't full proof, however. Physicist Frank Wilczek tells us the Standard Model of Particle Physics is flawed; its equations are lopsided and contain several loosely connected pieces.[2] These models have provided us with a reasonable picture of the universe and how it works, but there are many missing pieces to the picture (*dark energy and dark matter*), and some areas are vague and a challenge to credulity (*missing anti-matter mystery*). These two models as wonderful as they are, cannot provide us with a complete and cohesive picture of the universe, and by themselves are unlikely sources for discovering an underlying reality.

General Relativity (GR) does a wonderful job of describing space, time, gravity, and the larger aspects of the universe. It predicted blackholes, gravitational waves, and time dilation, all of which have been confirmed many times. It's equations, however, only address these larger aspects of the universe and cannot be used to describe the micro-world of quantum particles. By itself it is not likely the source for a complete picture of reality, but its success certainly suggests it will have to be accounted for in such a picture.

Quantum Theory (QT) accurately describes particles, fields, and the smallest aspects of the universe, but there is much controversy over whether QT represents an underlying reality. Physicist Niels Bohr and other architects of quantum mechanics believed it to be a good tool for describing/measuring aspects of particles, but they did not believe it should be taken as a broader theory as to what is real.[3] Einstein was uncomfortable with quantum theory, and over the years many others have felt discomfort in the disparity between the world as we see it and the accepted model of the quantum world where nothing is really what it seems.[4] Physicist Michio Kaku says, 'Quantum Theory is the most ridiculous theory ever proposed in the history of science.'[5] But, as he points out, it works and has withstood every experimental test hurled at it. Still, it is difficult to believe that a theory believed to be the most ridiculous theory ever proposed would at the same time form a sound basis for describing all reality. QT accurately describes our micro world quite well mathematically, but its interpretation of that world leads to some unusual conclusions, often referred to as quantum weirdness. Can such weirdness serve as a basis for our reality? We need to examine Quantum Theory a little closer.

Quantum Theory (QT)

The development of QT is an interesting story that starts well before Isaac Newton, when fundamental matter was theorized to be made up of tiny, physical objects called atoms. In the seventeenth century Robert Hooke and Christian Huygens developed a wave theory of light, while Newton had developed a 'corpuscular' theory of light, both of which contained reasonable evidence for their respective theories. Later developments confirmed light to theoretically having both wave and corpuscular characteristics.[6]

With Einstein's 1905 paper on the photoelectric effect, light energy was proven to be particles (*photons*), thus giving them definite particle-like *and* wave-like characteristics. Physicist Louis de Broglie later advanced a theory that all particles have wave-like attributes, which proved accurate. The recognition that two related particle attributes could not be precisely measured simultaneously gave rise to the Heisenberg Uncertainty Principle (*discussed shortly*). That in turn gave rise to the concept of superposition, wherein a quantum object could be in two states at the same time.

All these observations and developments resulted in the need for a new mathematical structure that recognized that experimental outcomes involving quantum particles was not certain and subject only to probability. With the development of Schrodinger's equation containing a *wave-function* to represent the wave-like potential and probability for particle measurements, Quantum Mechanics was born. All this brilliant work was accomplished before 1930. It has been refined and extrapolated since then, for the most part making things even weirder.

There were different schools of thought on how to interpret quantum mechanics. One school was to make no interpretation at all: the mathematics of quantum mechanics works very well so don't bother with interpreting its meaning. That school of thought was known as the Copenhagen Interpretation, which was apparently influenced by Logical Positivism, a popular philosophy at the time. That philosophy was founded on the idea that the only things we can know is what we observe, and what we are not observing does not exist, or at least asking questions about it is senseless. That particles don't exist if we are not observing them, and don't have attributes until we measure them, is called a non-realist philosophy.

According to the Copenhagen group, quantum mechanics tells us nothing whatsoever about the world. It is merely a tool for calculating the probabilities of various outcomes of experiments.[7] Due to the strong character of its primary proponent, Swedish Physicist Niels Bohr, that interpretation dominated for many years and is still taught in universities today. One of the reasons it has sustained over the years is that there is no consensus on a better interpretation of Quantum Mechanics.

Einstein was a realist. He believed particles known to exist continue to exist whether observed or not, and that particle attributes exist both before and after measurement. For many years he carried on verbal battles with Niels Bohr who of course believed that any discussion about unobserved particles and their attributes was senseless. Bohr was perfectly content with both quantum duality and quantum uncertainty and did not believe an underlying meaning was necessary. His philosophy was summarized with the expression, 'Shut up and calculate.'

An attempt to put quantum mechanics back into the 'realist' philosophy camp was put forth by Louis de Broglie, and later by Physicist David Bohm, called the *pilot-wave theory* (*discussed shortly*), but it was never accepted as a viable theory.

Another effort to put quantum theory in the realist camp was put forth by Physicist Hugh Everett, who suggested superposition was not an issue of certainty and uncertainty and that all states could be real if the universe *split* at each superposition. In this way both states are real, they each being in separate universes with different realities. This notion became known as the many-worlds theory.[8] While the many-worlds theory might be a usable interpretation of quantum weirdness, it is also unprovable, so it stays on the fringe of viable scientific theories.

Quantum mechanics has proven to be reasonably accurate and is therefore believed to represent a certain reality of nature. But it is fraught with multiple levels of quantum weirdness. Let's examine that quantum weirdness more closely, starting with the idea of uncertainty.

Uncertainty

Science is often looked upon as the search for the truth about nature. For physicists, this means finding the truth about matter: how it moves and interacts, and how these things change with time. Newton's observations and mathematics described things pretty well. Once scientists began exploring below the atomic level, however, the movement and interactions that changed over time were not so easily explained with Newton's laws. The belief that particles could be both point-like objects and waves, and exist in two distinct states, created a duality difficult to explain, making the quantum world appear quite different from the macro-world we live in.

Quantum Theory deals in the realm of subatomic particles, which cannot be seen, but are inferred from experimental evidence. Because we can't see the particles, we create sensitive equipment to detect them, but even that equipment has its limits. Trying to measure the characteristics of tiny particles we can't see and can barely detect is a challenge. Comparatively, we are still quite distant from such things as quarks and gluons. Imagine you are a scientist trying to conduct experiments by observing people on earth from as far away as the International Space Station (ISS). Through a special apparatus one could gain an idea of what the subjects on earth were doing, but given the distance and sizes involved, one could never be certain of knowing their precise movements or the causes affecting their movements.

Bouncing something as light as a photon off a particle to gain some information about it can change the particle's movement or characteristics.[9] Even accounting for the bouncing effects, we can only obtain limited knowledge from a single encounter with the particle; if we determine its exact location, we cannot know its exact momentum and vice versa.[10]

Looking deeper and deeper into the makeup of matter, it was inevitable that we would reach a limit to what we could observe or measure given the size and speed of subatomic particles. This leaves us *uncertain* when trying to obtain two characteristics of a particle simultaneously. That limitation is encapsulated in the *Heisenberg Uncertainty Principle* (HUP), which is based on the underlying notion that objective observations of nature's most basic entities are impossible.[11]

The uncertainty principle, developed by physicist Werner Heisenberg in the 1920s, stimulated the development of Quantum Mechanics, the mathematical formulation that gives probable answers when dealing with particles and their interactions. Quantum mechanics has since broadened into Quantum Theory that encompasses all aspects of matter and energy at the quantum (*most tiny*) level. Being so important to the foundation of quantum theory, we should take a closer look at this important uncertainty principle.

The Heisenberg Uncertainty Principle (HUP)

In Quantum Theory, the HUP is a fundamental principle that explains why it is impossible to simultaneously measure more than one quantum variable, such as position and momentum (*think speed*), or energy content and time. This principle was formulated when Heisenberg was trying to build an intuitive model of quantum physics. He discovered that there were certain fundamental factors that limited our actions in knowing certain quantities. He created a core equation that accurately expresses this limitation; the more we know about one variable, the less we can know about the other.

For a particle to remain a 'particle' over time [*as opposed to a wave*], it must possess definite values of position and velocity, but the Heisenberg Uncertainty Principle denies this possibility, giving rise to a delocalized wave-like aspect of quantum particles.[12] In other words, since a particle cannot possess both a specific position and momentum, it possesses a wide range of both values at the same time.[13] These theoretical ideas are what makes the quantum world appear weird.

One interpretation of the HUP is to recognize it as a purely mathematical deduction. Taking a measurement is a human endeavor. If when doing so we cause a particle's momentum to change, it should not be interpreted as a reflection on the particle itself. If we avoid disturbing the target particle and find we cannot then determine its precise momentum, we should not interpret *that* as a reflection on the particle either. In other words, our inability to measure does not justify assigning variable characteristics to particles.

The HUP is an important and necessary principle that needs to be taken into consideration when predicting experimental outcomes, but the principle is associated with a human endeavor and should not be applied to the description or behavior of particles independent of that endeavor. From this line of thinking the appropriate interpretation of the HUP is that it is a very helpful mathematical deduction that simply expresses a human limitation in simultaneously measuring two self-created particle attributes.

There are, of course, opposite opinions. Some physicists have adopted the HUP as representative of a fundamental aspect of particles themselves. If the position of a moving object is known but not its velocity, in the next instant of time, having a range of velocities available to it, the object would effectively come to occupy a range of different positions simultaneously. In other words, it would become a wave of superimposed position states.[14]

This interpretation says that objects subject to the HUP no longer occupy a specific position when their velocities are known and have no specific velocity when their position is known. When we know a particle's position, then it is behaving like a particle. But it does so only instantaneously. As soon as we look away, the HUP dictates that the particle must start behaving like a wave again because its velocity is not well determined.[15] The idea of multiple states is called *superposition.* It should be noted that this interpretation requires a belief in wave-particle duality (*discussed shortly*), without which there would be no basis for the idea of superposition.

It seems more reasonable that the HUP simply expresses a human limitation and says nothing about the attributes of particles themselves. The limitation inherent in the HUP makes sense when one realizes two things. Firstly, one of the two quantum variables under consideration is in constant change. Momentum is defined as velocity times mass. Velocity means the particle under consideration is moving, i.e., its position is constantly changing.

Secondly, one must realize that there is no such thing as "now". Now is a term we use generally to refer to that point in time between past and future; it falls just after the past and just before the future. The nature of time precludes us from ever experiencing or measuring a specific "now" because before one can recognize and record it, it has already passed. Consequently, if a particle is constantly changing position due to it experiencing momentum, and there is no way to record "now" during that movement, there is no way to know with absolute precision the particle's momentum *and* position simultaneously.

One could pick a future time and say that if conditions don't change, meaning the rate of change remains constant, then one could *predict* a measurement value. This is in essence what quantum mechanics does, it predicts a value given a calculable rate of change.

Understanding Uncertainty

A thought experiment might provide some clarity as to why precise simultaneous measurements are seemingly impossible. A fan with a movie camera at a baseball game films a homerun hit over the center field fence and later she wishes to measure the ball's velocity with respect to its location. She knows the directional aspect of its velocity, pinpointed by a straight line between home plate and where the ball crossed the center field fence. From a single frame of the movie film, she can produce a still photograph that pinpoints the ball's location, but because the ball is not moving in the photograph, she cannot determine the ball's speed.

Measuring the distance the ball travels between several movie frames and knowing the timing of the frames, she can calculate the ball's average speed between those frames. Unfortunately, she cannot know exactly where the ball was in those frames when that speed was achieved. She can narrow the number of frames down making the distance between them less and less. Even measuring the distance and time between just two frames, she will only know the average speed and not know where in that distance, no matter how small, the ball was when it achieved that speed.

She can get very close to knowing the ball's speed at a specific point, limited by the filming capacity of her camera, but no matter how good of camera she uses she will never be able to calculate both the speed and location *precisely.* On the scale of a particle's movement the methods of calculating its location and momentum may be quite different, but I believe this gross description of the process gives the reader an idea of what physicists are up against trying to measure two aspects simultaneously.

Let's explore a little deeper why we may have such limitations and uncertainties requiring a math based on probability rather than certainty. If the only method we possess to determine a particle's location and momentum is to bounce other particles off it, which may alter the target particle's characteristics, then understandably we must resort to probabilities rather than certainty. We may someday develop a different method to ascertain this information, or maybe we will find that learning both simultaneously is actually impossible.

We may never be able to truly know or understand the minute environment in which these tiny, unseen particles reside. We know the environment is crowded and contains many different conditions, and there might even be more layers of conditions we have not yet detected. Let's take a closer look at the environment in the science lab where a test subject viewed from the ISS is under investigation to determine its location and momentum simultaneously.

The Minute Environment of a Subatomic Particle

Scientists once believed the 'vacuum' of space was nearly empty and was very much like a vacuum, but now they believe it is teeming with activity. Physicist Harold Fritzsch tells us that up close the vacuum looks more like a bubbling cauldron than empty space.[16] Physicist Kerson Huang tells us the vacuum's fluctuating electromagnetic field buffets an electron as if in a stormy sea.[17] If the vacuum is that raging, then the environment around the science lab must be a *really* turbulent place, be it in either Chicago or the International Space Station.

First, the medium in the lab is bombarded by a very high density of particles. One-hundred trillion neutrinos pass through our bodies every second.[18] We know that huge quantities of near ubiquitous photons are likewise bombarding us. The medium might also contain a high density of dark matter, the most dominant source of matter in the universe.[19] Negotiating through this blizzard of particles would be like negotiating a New York City sidewalk at 5:00 pm on a Friday. One would be challenged to maintain a straight path and constant speed with so much getting in the way.

Typical of New York City, it is bathed in a rainstorm, so the umbrellas are out. The umbrella represents a particle's field. It takes up space around all test subjects, so it is something else to either avoid or respond to as the sidewalk is negotiated, causing one to flit about relative to one's own umbrella (*field*). The fields are coming from all directions, and some (*electric fields*) are rather rude in their pushing and pulling, giving one at times a staggering gait. Occasionally one must absorb or emit energy packets (*photons*), altering one's momentum and challenging one's further effort to stay the course.

Then the weather turns severe. Cosmic rays (*high energy space-borne particles*) and very high energy gamma rays (*photons*) are constantly raining down at high speed. Gravitational waves of various degrees are constantly rolling through. Penergy fluctuations cause virtual particles to snap in and out of existence all around. All of this distorts the environment, making it much more like a *bubbling cauldron.*

Negotiating a straight path with constant speed under these conditions is seemingly impossible. It is no wonder a scientist on the ISS sees his test subject moving in a straight line with constant speed, but from the perspective on the sidewalk that is hardly the case. Constantly reacting to the onslaught of particles, fields, waves, forces, and energy fluctuations, one must flit about so quickly within one's field it seems like one must sometimes be in two places at once. The best the satellite scientist can do is wonder why a test subject seems to be in two states, and why the scientist can only give a probability of where it will be and how fast it will be traveling at any future time.

The above analogy is not exactly a description of the subatomic environment that exists on earth, but it does demonstrate how busy the Penergy medium is and how influential and important fields are. Some things we should take away from this scenario are: (1) Particles have a field that surrounds the particle like a bubble or cocoon; (2) Particles move around within that cocoon as demanded by the conditions in which they encounter other particles, fields, waves, and energy conditions; (3) We have no way of sensing all that is going on within the particle's minute environment; and, (4) We are left with only the ability to calculate a probability of an outcome when trying to determine such things as a particle's precise location, its scattering direction, its likely decay process, its exact energy, its tunneling capabilities, and other attributes of particles and their behavior. The minute world of the particle is difficult to measure and seemingly weird, but perhaps understandably so.

A premise of the Tor Model is that an individual particle resides within its own individual bubble-like cocoon we call its field. What is the theory for the existence of such a cocoon?

A Particle's Personal Field

In the Tor Model, fields are not ubiquitous throughout the universe, but are personal to individual particles and groups of particles making up any size body. A bubble-like cocoon is of course used only as a visual image for the presence of a field surrounding the particle. There is no actual, physical bubble or cocoon present. We can't see individual particle fields only their effects, but we know enough about how they affect other particles to have a vision of their presence. There is no dispute as to particles having fields, only the source of those fields. The argument for a particle having a local, bubble-like field is found in our understanding of particle location, which is worth examining.

According to quantum theory as told by Tim James in his book *Fundamental*, we cannot know a particle's precise location until we measure it, and that particle locations are determined by waves of probability.[20] He points out that particles can be pushed through the same experiment over and over and come out in a different place each time. Quantum Theory ascribes this phenomenon to the theory the particle is also a wave, which gives it an indeterminate location. Given the current belief in particle duality, this notion is justified, but not necessarily accurate.

Scientists never see particles as waves, only as point-like particles.[21] This is important so allow me to repeat it. *Scientists never see particles as waves, only as point-like particles.* In other words, there is no physical evidence a particle is also a wave. Particles are believed to be waves based on the interpretation of experimental outcomes, such as in the double-slit experiment (*discussed in the next section*) and from the experience of a repeated experiment resulting in different location outcomes. These experimental outcomes seem counterintuitive and weird, fostering a wave-like interpretation, but they can be explained by visualizing a particle that is producing its own individual, local field within the Penergy medium.

In the Tor Model, a particle is not a wave. It can seem to behave like a wave because its field, created by the particle's spinning gyrations disturbing the Penergy medium, has the characteristics of a wave. As James Geach tells us in his recent book, *Five Photons*, '... for a charged particle surrounded by an electrostatic field, any oscillation of the particle will also oscillate the field.'[22] Consequently, as the particle moves, it constantly affects the Penergy medium creating a bubble of disturbed Penergy around itself, which becomes the particle's individual field.

The bubble, or cocoon of disturbed Penergy is not fixed. It is very fluid, and its dimensions, shape, and behavior are all subject to constant change depending on what the particle is responding to at any moment. The bubble can be distorted or even collapsed depending on what the particle encounters. Because a particle is constantly affecting the Penergy medium, it is in essence constantly creating a new bubble of disturbed Penergy around itself as it moves. The on-going creation of the bubble causes a lag time between the location of the particle and the location of the border of its field. The bubble is constantly shifting around the particle as the particle is moving about within its *minute environment* as described above.

The particle's gyrations create vibrations that change the state of the Penergy, creating a distinct environment that defines the bubble. Within that bubble environment the particle is free to roam while at the same time being somewhat confined by the bubble. While the particle is creating the bubble, the bubble is giving the particle a safe and defined place in which to reside. It is somewhat like a barrier protecting the particle; a first line of defense when encountering another particle. The practical result is that the particle can be found most anywhere within its bubble-field, with its precise location following mathematical probabilities. Those probabilities are accurate using a wave-like theory because the particle's movements, its field, and the Penergy medium are all behaving in a wave-like manner.

Given the above description of a particle's personal field and the idea that some fields are without boundaries, it is understandable why quantum theory posits particles could be anywhere until they are measured or detected. Some fields such as the electric field may be without a boundary technically, but the density of the field falls off so quickly that its practical boundary, where it has an insignificant influence on other particles, is not terribly far from the particle itself. Therefore, the probability of finding the particle far away from the center of its field becomes too remote to even consider. Again, at any moment the particle might be found anywhere within its field, which explains why one would obtain different location-results with an identically repeated experiment.

Another instance of particle behavior suggesting it is a wave has to do with its apparent capacity to tunnel into places to which it theoretically should be incapable of moving. A particle shot at the edge of a wall will usually bounce back, as we would expect a tennis ball to do. But occasionally the particle can be experimentally found on the other side of the wall.[23] This is known as particle tunneling.

In quantum theory the interpretation of the tunneling phenomena is that the particle is a wave that occasionally diffracts (*bends*) around the edge of the wall, as waves are known to do. Its bending around the edge of the wall allows the particle to show up on the other side. This is of course only possible if the particle is a wave.

In the Tor Model it is the particle's field that is diffracting around the edge of the wall. Since the particle is seldom located at the edge of its field, it is seldom that it is within the portion of the field that is being diffracted so it seldom appears on the other side, but occasionally it is at the edge of its field and is diffracted around the corner. We can learn more about wave-particle duality and diffraction in the famous double-slit experiment discussed next.

> Deduction #12: Much of quantum weirdness (*particle phenomena without explanation*), and an explanation for why a particle's precise location and behavior is subject only to probability, can be explained by understanding how a particle moves about within its local, personal field.

The Double-Slit Experiment

One of the cornerstones of Quantum Theory is the idea that particles are both a point-like particle and a wave, giving their behavior a great deal of uncertainty and possibly meaning particles are not real at all. True enough, particles exhibit point-like particle attributes in scattering experiments where they ricochet off each other. And they exhibit wave-like attributes in experiments like those involving electron beams striking crystalline solids in which the electrons exhibit diffraction and interference, both attributes found only in waves.[24] But perhaps there is a more plausible explanation for this phenomenon.

Quantum objects [particles] can appear to be both point-like or wave-like depending on how one has chosen to detect or measure them.[25] For physicists the proof of the wave-particle duality comes in the way of the *double-slit experiment*. When a wave is washed against a thin wall having two closely placed vertical slits, the slits will cause the passing wave to diffract (*bend around the edges*) as it passes through the slits. The semicircular waves emerging from the two slits will interact with each other because the wavefronts will start to cross each other's paths. The diffracted waves from the two slits interfere with each other both constructively and destructively, creating crests and troughs that show up vividly on a detection screen set just beyond the wall. This setup is known as the double-slit experiment. For a complete description of the experiment, see Philip Ball's *Beyond Weird*.[26]

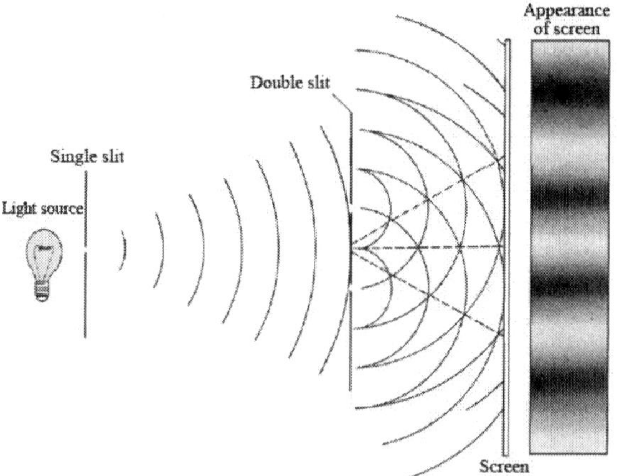

Figure 7.1 Graphic of the double-slit experiment. (Source: Penn State University Physics)

When sending either photons or electrons through a two-slit setup the detection screen will show the wave like pattern of crests and troughs, as shown on the right-hand side of Figure 7.1. The conclusion drawn from these experiments is that the individual particles are passing through *both slits at once*, interfering with themselves and therefore are not only acting like waves, but *are* waves. They originate as particles, and end as particles at the detection screen, but in between they act like waves.[27]

The conclusion being drawn is understandable, but it might not be correct. One can also imagine the same experimental outcome with each particle passing through only one slit but their *fields* passing through both slits. It might be the field that is being divided and diffracted. The wave-like appearance of particle locations on the detection screen would be due to the particle itself being somewhere inside the divided and diffracted field. The particle would strike the screen before the two fields could recombine. Its exact location would be consistent with the probability of finding it greater in the constructive waves and lesser in the diminutive destructive waves, hence the detected interference pattern.

It is also noted that in a separate experiment if you place a detection device at each slit to determine which slit each electron actually went through, the interference pattern does not show up on the detection screen.[28] The reason for this may be that the detection device caused the particle's field to be significantly disturbed and effectively collapse at the entry to the slit, thus the field did not go through both slits and therefore could not produce the interference pattern. This too suggests that in the first experiment it is the particle's field, and not the particle, that is passing through both slits and causing the interference pattern.

Particle Field Collapse

Let's examine what is meant by particle field collapse. Recall that in the Tor Model, the spinning undulations of the particle disturbs the surrounding Penergy medium creating a tension in the medium that defines the particle's field. As long as the particle is spinning in the Penergy medium it will always produce a field, but that field does not have a precise boundary. In circumstances where two or more particle fields are forced together, the fields of multiple particles could be compelled to momentarily overlap.

Because the different particles have different tension values, spin rates, and other spin characteristics, the overlapping of the fields creates an area of turbulence that effectively collapses the usability of that portion of the field. A detection device at each slit could be the cause of the fields passing through the slits being crushed or otherwise disturbed to the point of creating such a turbulence that effectively collapses much of the field. Consequently, the two halves of the field cannot effectively interfere with each other, and no interference pattern is displayed on the detection screen.

The Pilot-Wave Theory

Again, particles are always observed as point-like particles, never as waves. In the Tor Model this is because they *are* only point-like particles, but their existence inside the bubble-like cocoon of their pulsating field makes them behave like waves. Physicist Louis de Broglie in the 1920's, and David Bohm in the 1950's, appear to have nearly figured that out. They advanced a theory that point-like particles are guided by, or riding on, *pilot-waves.* The pilot-wave theory describes everything quantum mechanics does and explains a good deal more.[29] The theory assures that particles are both real and present whether observed or not, in opposition to then popular arguments to the contrary.[30]

Bohm believed particles are being pushed around by these invisible guide waves, thus would appear to move in wave-like trajectories.[31] The photons in the double-slit experiment were likewise surfing on pilot waves and it was the pilot wave passing through both slits that caused the interference pattern on the detection screen.[32] Evidently Bohm's mathematics was good, deviating little from current quantum physics, but he could never prove the existence of pilot waves. Initial experiments looked promising but could not be duplicated. The de Broglie-Bohm theory retains a small following, but the theory remains outside the mainstream of quantum physics.[33]

Bohm thought the pilot wave was a vibration in some pervasive and sensitive field he called the *quantum potential*. That description sounds very much like the medium we are calling Penergy. According to Physicist Philip Ball there is nothing obviously impossible about the idea of a quantum potential, it's just that there is no evidence for it.[34] It is quite possible the quantum potential Bohm had in mind was in fact our Penergy.

A special prize may await the young physicist who extrapolates Bohm's mathematics for pilot waves into the physics of particle fields, which may explain the double-slit interference pattern accordingly and resolve the issue of wave-particle duality once and for all. That prize might be shared with the equally enterprising young physicist that devises and carries out the experiment that proves it is the particle's field that is being diffracted and passing through both slits.

Particle Entanglement

Another demonstration of particle peculiarity necessitating Quantum Theory is the topic of *entanglement*. If two like particles originate from the same source or otherwise meet, they are said to be entangled, meaning they somehow have the means to communicate over significant distances without anything passing between them. A classic experiment demonstrating entanglement involves photon polarization.[35] Polarization is the term used to describe the direction a photon is waving as it travels. We can't tell what direction the wave is pointing until passing the photon through a polarizer set at a certain direction. If it passes through the polarizer, we know that its wave was pointing in the same direction at which the polarizer was set.

In the basic experiment if two photons originate from the same source and are then sent out in opposite directions toward two polarizers, the photons will do the same thing in terms of passing through the equally set polarizers. If one photon goes through a polarizer, the other photon will do the same. If the first photon is blocked at the polarizer, the other photon will also be blocked.[36] How does the second photon know at what angle the first photon was traveling in order to mimic its behavior? The presumption is that they somehow communicated that information between them just before the second polarizer was encountered.

Similar experiments have been carried out using the spin-up and spin-down characteristics of electrons, and in all cases it appears the electrons were somehow communicating with each other to statistically arrive at the correlation predicted by quantum mechanics. These experimental outcomes bothered many physicists including Einstein, who didn't like the idea of one particle affecting another without anything passing between them. He called it *spooky action at a distance.*

Entanglement allowing communication between distant particles without anything passing between them is based on some assumptions. It assumes the particles do not have any definite properties prior to being observed or measured;[37] that the properties of a particle aren't fixed until we measure them;[38] and that the polarization angle cannot be preset by the photons before starting out.[39] These are big assumptions. They are based largely on the non-realist approach consistent with the previously mentioned Copenhagen Interpretation.

Einstein believed particles could have such properties, which became known as *hidden variables.* A mathematical theorem created by physicist John Bell in 1964 based on statistics and the Uncertainty Principle is said to prove hidden variables cannot be experimentally verifiable. All entanglement experiments to date have been consistent with Bell's theorem, supporting the idea hidden variables cannot be verified and likely do not exist. Experiments have never found hidden variables, but that doesn't mean they don't exist, just that they have not yet been found as defined.

Whether unobserved properties known as hidden variables exist, as Einstein believed notwithstanding Bell's Theorem, is still neither proven nor disproven so their existence remains an open question.[40] In the Tor Model some of those hidden variables may well exist residing in the particle's field, as discussed in Chapter Five.

The idea of entanglement is being used in the development of cryptography and quantum computing and many experiments point to it being very real. Yet, perhaps there is more for us to understand. Undoubtedly scientists are finding two entangled photons miles apart having the expected properties, but when and how they acquired those properties is still an open question. It could be that when the two particles touched, they synchronized the variables in their respective *fields.* Entanglement could turn out to be a good argument for both particles producing fields and particles communicating through those fields. More experimentation is necessary, and we must assure we have a full understanding of particle fields and all their aspects and potentialities before drawing final conclusions.

It is possible that our thin Penergy medium is a very dynamic form of energy capable of much more than we can possibly even imagine. It could be the means for a common inter-connectedness between all things and possibly even future cosmic communication. It is hoped that the scientific investigation into quantum entanglement will produce results suggesting how we might someday tap into that energy.

Notwithstanding all this quantum weirdness, quantum mechanics works very well. It consistently and accurately predicts experimental outcomes, though those predictions are expressed as probabilities. The original equation accounting for this ability is the Schrodinger equation containing the wavefunction. Let's examine the idea of the wavefunction and its effect on reality.

The Wavefunction

One of the issues in interpreting the meaning of quantum mechanics was a controversy regarding the meaning of one of its core equations, the Schrodinger equation. That equation contains a symbol called a *wavefunction.* It entails the notion the particle aspect under consideration behaves like a wave, with its measurement potentially having an infinite number of values that are evenly spread out in a wave pattern [*much like a field was described earlier*].[41] When a measurement is taken, the wavefunction is said to collapse producing the probability value for the attribute in question.[42] The uncertainty about the meaning of the wavefunction is expressed in a controversy over interpreting when exactly the wavefunction collapses, whether it requires a conscious observer, or even if it collapses at all.[43] That issue is known as the *measurement problem.*

The wavefunction is a mathematical term, not a real thing. No experiment has ever proven the existence of a wavefunction.[44] Quantum mechanics is accurate and is used widely throughout science and industry, consequently many see the wavefunction as something not only very meaningful, but very real. It has been elevated to the level of being part of reality, and for some it is, 'the sum total of reality'.[45] Now in quantum physics everything has a wavefunction; your book, the earth, even the whole universe has a wavefunction.[46] An object's wavefunction determines its behavior and the behavior of an object's wavefunction is determined in turn by the Schrodinger equation.[47] Apparently for those mathematically oriented scientists, the world and everything in it has been reduced to the status of mere elements of an equation.

Again, quantum mechanics is beautiful and accurate, but its success should not be elevated above its purpose. It being capable of giving accurate probability answers is not so mysterious. The experimental outcome determining a particle's attribute takes place when a measurement is taken and the wavefunction collapses. In the Tor Model, much of the wavefunction's success and accuracy can be ascribed to the wavefunction mimicking the particle's wavy field. To measure a particle's attribute, one must interfere with the particle, which collapses its field. When the wavefunction collapses to reveal the value of a particle's physical attribute, so does the field around the particle collapse, exposing the particle's attribute. The collapse of the wave function rhyming with the collapse of the particle's field is what makes QM seemingly so statistically accurate.

We have examined quantum uncertainty, quantum superposition, wave-particle duality, quantum entanglement, and the wave-function of quantum mechanics. It is time for an assessment of whether quantum theory can be used as a basis for reality.

The Reality of Quantum Theory

Reality is the state of things as they actually exist, as opposed to an idealistic or purely mathematical idea of them. We once thought reality was just as we saw the world, but we are told that what we see is only on the surface of a deeper reality, a deeper existence. That deeper existence is described by quantum theory in terms of such things as a dual nature of particles of matter, the possibility nothing exists but fields, or the possibility we live in multiple worlds. All these notions suggest a reality much different than the one we witness daily. The enigma at the heart of quantum reality can be summed up in the motto: What we see when we look at the world seems to be fundamentally different from what actually is.[48] Given the weirdness of these quantum ideas, can quantum theory provide us with a valid picture of reality?

There is no question quantum theory math works. The only question is whether the interpretation of quantum theory represents the underlying reality of how the universe is. Some of the weirdness of Quantum theory's picture of reality is self-induced. It has adopted extrapolations of the HUP such as *physical systems do not possess attributes until those attributes are measured.* Physicist Yasunori Nomura characterizes this assertion as a metaphysical claim.[49] That means that it is not scientifically proven. It seems to be a concept left over from the Copenhagen Interpretation that was at the heart of the controversy between Niels Bohr and Einstein. Einstein of course believed that particles do have definite values for their position and momentum, and just because we can't determine them simultaneously doesn't mean they don't exist or that measuring only one allows the other to take on numerous values.

Another quantum theory extrapolation of the HUP is that if you know an electron's momentum you cannot know its location, and if you cannot know exactly where it is then it exists in several *parallel states simultaneously.*[50] The Tor Model's answer to this is that you don't know the electron's precise location because it is flitting about within its bubble-like field. In Quantum Theory the electron can be flitting about anywhere in the universe, which is an extrapolation gone too far, creating its own weirdness.

Einstein believed that quantum theory, though giving correct results was incomplete and he did not believe a wavefunction could be used to describe reality.[51] Other physicists accept quantum theory as very much describing reality. They have extrapolated the math of the HUP and quantum mechanics into a broad spectrum of ideas including time does not exist,[52] nothing is real[53], particles don't exist but their fields do[54], very little can be known with certainty, and we probably exist in one of many worlds with similar but different histories.[55] An excellent account of all these theories can be found in Sean Carroll's, *Something Deeply Hidden*.

These ideas are principally math driven. Although math can produce accurate experimental results, it is a poor basis for being used to describe reality. Tim James tells us quantum mechanics is a triumph of mathematical beauty, provided you don't ask what any of it means.[56] Physicist Sean Carroll tells us that the things math proves is not true facts about the world[57], and that physicists don't know what Quantum Theory actually means.[58] The elements of quantum theory though accurate mathematically, may not produce a usable picture of reality for the following reasons.

- Entanglement is a very interesting phenomenon, but it may be no more than that. Experiments have proven that particles have attributes, some of which are fixed and some that are subject to variation such as the direction of its spin or polarization. Experiments have also proven that two particles when entangled by touching or having emanated from the same source can share those attributes in statistically predictable ways. Quantum Theory says the statistical predictability is due to the particles having the capacity to communicate rapidly that allows the second particle to adjust its attribute depending on the value of the first particle's attribute when measured. When and how each particle acquired the value of the attribute and whether they can communicate or preset those variables is still an open question. Particle communication without anything passing between them is a respectable theory, but it is inconclusive and cannot yet be considered a part of our reality.

- Wave-particle duality may not be real. If future experiments demonstrate that the waviness of a particle is a trait inherent in its personal field and that it is the field that passes through the double-slit experiment causing the interference pattern, then the wave-particle duality issue might be put to rest. Wave-particle duality therefore would no longer be part of our reality.
- If particles are not doing the waving, it means they have definite positions. It also means superpositions are dubious, and that a particle seemingly in multiple locations may be explained by it flitting about within its own personal bubble-like field as explained above. Superposition therefore may not be part of our reality.
- Schrodinger's wave-function equation provides accurate experimental predictions. The validity of the wave-function calculations, however, could be due to the wavefunction mimicking the wave-like attributes of the particle's field as explained above. The wave-function therefore remains accurate but describes a reality much different from the reality interpreted by Quantum Theory.

- The HUP is an important concept and mathematical tool but is perhaps nothing more. Without the concepts of particle duality there is no support for superposition and the HUP becomes no more than a human limitation to measurement.

Wave-particle duality, the HUP, superposition, and the wave-function predicting accurate probabilities, have all been brilliant answers to real, head-scratching problems. Those answers however, created a picture of quantum weirdness that may have led us down mathematical roads to realities that don't exist. Imagining a reality that stems from the belief the micro-world has a dual nature and works strictly based on probability is perilous; fraught with opportunities for miss interpretation.

Lee Smolin tells us that the use of probabilities is just for our convenience and the resulting uncertainties just an expression of our ignorance.[59] Physicist Kenneth Ford points out that the use of probability in the larger world results when we don't have all the available facts.[60] That might be the case for Quantum Theory as well; we just don't yet have the means to know everything that is going on in a particle's minute environment. Physicist Yasunori Nomura tells us we may never really know what is going on in the quantum world.[61]

The goal of physics is to understand the laws of nature in order to anticipate and describe the future.[62] We strive for that goal, but we may never have a complete theory or mathematical structure to precisely describe the minute environment of subatomic particles. Consequently, we may never understand all the laws of nature that exist for them or develop a reality that includes precisely all the aspects of that micro-world.

So where does that leave us in search of a reality? We may have to go back to our slate of observations and deductions and see what it can tell us.

We know that pure energy (Penergy) existed at the beginning of the universe and is the substance from which everything in the universe is made. It apparently can exist in a range of densities, which explains the creation of supermassive blackholes, the origin of the cosmic web, how particles were created, and why the universe appears so homogeneous. The existence of Penergy in a constantly thinning density may be the source of both our interstellar medium and the dark energy that has been expanding the universe since its origin.

The Penergy medium may be a better backdrop than our current theory of undetectable, ubiquitous fields for each type of particle. If all that is true, then particles are real and their fields are very likely simply a disturbance in the Penergy medium. It means our reality will come from a realist point of view as Einstein insisted. If the Tor Model's interpretation of the double-slit experiment is correct, it could mean QT could be interpreted as a realist theory, and perhaps the combination of a reinterpreted Quantum Theory and General Relativity is still a viable source for reality

The Reality of Reality

It is hoped that finding a way to combine General Relativity (GR) and Quantum Theory (QT) will provide a single theory of everything and expose the underlying reality if one exists. They differ in some important respects, so bringing them together has been a real challenge. They have deeply different descriptions of time. QT has a single universal time. GR has many times depending on one's relative motion and relative position in a gravitational field.[63] In QT things can be in superposition states. GR does not have a superposition principle.[64] However, given the above analysis, this may be a moot point.

The combination of GR and QT might someday be possible, but that combination may still only describe the mechanics of the universe, perhaps exposing some additional hidden elements, but that will still not likely provide a picture of the larger reality. Even together they don't seem to encompass everything going on in the universe. As Physicist Lisa Randall points out, the combination of GR and QT into a single set of equations that describes particles and their interactions might be possible, but they won't describe reality.[65]

Does this mean that none of the monuments to human thought, exploration, ingenuity, discovery, and creativity have produced a realistic picture of reality? Have we missed seeing the forest through the trees? That might well be the case.

In the Tor Model, to develop a true picture of the universe one must start not with QT and GR, both of which are excellent tools for measuring aspects of the universe, but instead start with the dynamics of pure energy, the substance from which the entire universe is made. Penergy and the changes it has undergone is not only the basis of cosmic history, but it could also be the basis of cosmic reality as well. The dynamics of Penergy of course include the work of the CAGI, the ultimate source of cosmic change. Their work together can be encapsulated in a few lines.

Penergy produces matter.

Matter produces fields.

Fields produce forces.

Forces produce Levels of complex matter, including life.

Life produces intelligence and technology.

Intelligence & technology produce greater complexity.

Greater complexity begets greater complexity.

This is the substance and reality of the evolution of our universe. There is more to come and more lines to add. The changes at each Level are becoming greater and the emergent qualities at each Level more wondrous. From this vantage point reality is much bigger than the mechanics working at either the micro or macro level of existence.

This sequence has not yet been reduced to mathematical terms, but it seems to be a pattern that lends itself to such a formulation, and perhaps a schematic for a true theory of everything. It may provide a fairly accurate mathematical account of what has taken place, and perhaps will take place in the future. Such a formulation would be looking at the universe from a new and different perspective; one with perhaps a sharper focus.

Looking backward in time and deeper into the levels of matter does not seem to be giving us the picture of the reality we seek. We need to broaden our horizons, perhaps exploring a shift in our focus. A change in focus from the past to the future, and from the depths of matter to the future of matter may give us a better picture and a broader perspective. We will examine the prospect of a broader *perspective* in the next chapter.

Chapter Eight

New Perspectives

Our construction of the early universe is complete. Atoms have evolved, photons have been liberated, hydrogen and helium gases have formed, stars are in the making, and the SMBHs are pulling it all together to produce the beautiful galaxies that will continue to expand into the universe we see today. It has been quite a journey, with many unanticipated twists, turns, and surprising deductions. The journey has spurred many new ideas and theories, while at the same time challenging some old theories.

Our next question was whether all this construction has given us a new picture of reality, and we have realized that a picture of reality may well depend on our perspective. Things usually appear differently from different perspectives, but is there a certain perspective from which our view of reality is truer? Or is reality only the sum of views from all perspectives? In our further pursuit of a proper reality, let's examine some pertinent perspectives.

The Penergy Perspective

Pure energy, e.g., Penergy, has not been given its due. It is the most important substance in the universe yet is seldom mentioned and is then only referred to obliquely.

It has become synonymous with mass energy and is otherwise pretty much ignored. Scientists use the word energy often, but they are usually referring to one of the different forms of *work energy*, which is different than Penergy.

In the science community the definition of *energy* is the ability to do work. Work is the generic description for the transference of energy between various systems. Energy cannot be created or destroyed, but it can be transferred, though always with a loss, usually in the form of heat. Work energy exists in many forms, each of which is discussed below. All these forms have highly technical definitions, but for our purposes we need only a common, less-technical definition, taken largely from Wikipedia. To understand Penergy and work energy we must be clear about their relationship. Let's start by examining the different types of work energy.

Radiation: Radiation energy is that which travels as particles or waves, such as electromagnetic waves. Its physical manifestation is the photon. Radio waves, infrared waves, visible light waves, and x-rays are all radiation energy.

Light: Light energy is electromagnetic radiation in the visible range. Visible light is emitted within the frequency band capable of detection by the human eye. Light energy is emitted by any object holding heat, and since all objects hold some degree of heat, all objects emit radiation, including light energy.

Heat: While temperature is the measurement of heat contained in a body, heat energy, also called **thermal energy**, is the flow or transference of energy between two bodies. It is the exchange of photon energy by atoms or molecules bumping into each other, transferring energy from a higher energy (warmer) object to a lesser energy (cooler) object.

There are three ways of transferring heat energy: conduction, convection, and radiation. *Conduction* is the transfer of heat energy through a solid, as in leaving the tip of a hot poker in a fire and allowing the heat to gradually rise up the entire length of the poker or allowing the bottom of a pan to heat up by placing it on a burner. *Convection* transfers heat energy through a gas or liquid, as in the heating of a room or the boiling of water. *Radiation* is the transfer of heat energy through a space. It is that which one feels by standing in the sun or near an open fire. In all these cases, the electrons in the receiving body are gaining in energy.

Chemical: Chemical energy is the energy stored in the bonds of chemical compounds, normally carried in the electrons of the atom or molecule's outer rings. That energy may be released or absorbed during chemical reactions and is otherwise released in the form of heat (*radiation*).

Nuclear: Nuclear energy is a form of potential energy in the atomic nucleus. The energy is derived from the high energy required to form the nuclei and hold it together, and that energy is released during a nuclear reaction, especially during fission or fusion.

Kinetic: Kinetic energy is the energy a body possesses by virtue of its being in motion. The amount of measurable energy depends on the body's mass and velocity. The motion can be vibrational — such as in the plucking of a guitar string; transactional — the velocity gained or lost due to the absorption/emission of a photon, or translational — the gain/loss of velocity due to a collision with another particle or mass.

Gravitational: Bodies with mass are attracted to each other by their respective gravitational fields. Gravitational energy is the potential energy a body possesses due to its position that would allow gravity to influence its motion.

Potential: Potential energy is the energy possessed by a system due to the position of its parts. A body lifted from the earth's surface is said to have potential energy due to it being in a gravitational field and its potential to fall toward the earth's surface, whereby the potential energy would be converted to kinetic energy.

Common Traits

Let's look at these work energies a little closer to identify any common traits. *Heat* energy, whether transferred through a medium of a solid, liquid, or gas, comes down to a high-energy electron emitting a photon, which is received by a lower-energy electron, thus spreading the level of energy throughout the medium until an equilibrium is reached. Let's label this phenomenon of energy transferred through photons as **Photon Energy**. *Light* energy and *Radiation* energy are also Photon Energy, by definition. *Nuclear* energy is also Photon Energy because the energy released by the nucleus in fission, fusion, or decay is in the form of photons.

Gravitational energy is derived from a mass distorting spacetime, creating an attractive field by bringing the pathways around in toward the mass. Since it derives from the mass's field, let's label this phenomena **Field Energy**. *Chemical* energy is also Field Energy because it is based on the attractive/repulsive force of the electric field. That force affects the behavior of electrons in atomic/molecular interactions, which we call chemical reactions. *Potential* energy is also Field Energy because the *potential* for influencing the movement of a particle or mass is created by the presence of a field, either gravitational, magnetic, or electric. The potential energy is inherent in the field itself, which is derived from particle spin, a natural attribute of Penergy.

Kinetic energy is a hybrid between Field Energy and Photon Energy, as both can affect a particle's motion.

We've narrowed our work energies down to three categories; photon, field, and kinetic, which is a hybrid of both. Let's examine them a bit closer.

Photon Energy:

Photons are particles, but due to their configuration they present no mass, and due to their mimicking an electromagnetic field, they can be absorbed/emitted by matter particles. Photons are comprised of early elementary particles and do have substance, as evidenced by their possessing momentum. Photon momentum can knock electrons off the surface of the sheet of metal, as shown in Einstein's paper in 1905 on the Photoelectric Effect. Consequently, photons possess an energy that can be measured in mass energy and can be converted to Penergy.

Photons are the currency used by matter particles to exchange energy, providing the needed kinetic energy to a particle, atom, or molecule for it to attach to, decouple from, or otherwise interact with another particle, atom, or molecule. All these interactions are essential for matter to combine and grow at all levels.

When a matter particle absorbs a photon, it not only changes the velocity of the particle but the calculation for the particle's total energy, as well. The value for the new total energy is found in the calculation for the change in the particle's Kinetic energy, as discussed below.

Field Energy:

The other way to change a particle's velocity is for the particle to be acted upon by a force. In the Tor Model, forces are derived from gravitational, magnetic, and electric fields, all of which originate with the field or spin of the Tor particle. In all these fields an unincumbered particle will be caused to *accelerate*, meaning change its velocity, either its speed or its direction. The energy needed to change the particle's velocity is derived from the field. The value of that energy is also measured in the change of the particle's Kinetic energy. Field energy is also Potential energy.

Kinetic Energy:

The total energy (E) of a non-moving particle is tied up solely in its mass (M). Mass is ultimately a measure of, and equivalent to, its pure energy *(Penergy)* according to $ET=MC^2$. Once the particle is moving by the absorption of a photon or encounter with a force, the particle possesses additional energy in the form of Kinetic energy. The kinetic energy is a measure of its mass (M) and its velocity (V) according to $EK=.5MV^2$. So, the total energy of a moving mass is the sum of its Mass (*Penergy*) energy, and Kinetic (*Field/ Potential and Photon*) energy.

This is consistent with what Einstein taught us about energy; it takes energy to move a mass, and the more mass being moved and the faster it is traveling, the more energy it takes to change its speed or direction. This is especially so at speeds closer to light speed where relativistic effects come into play. In that case one would have to square each of the Mass and Kinetic energy values and take the square root of that number to arrive at the total energy value.[1]

Energy Summary:

The first role of Penergy in our universe was in the act of spinning itself into blackholes and particles. These two entities have the correct ratio of spin-rate-density to the surrounding-medium-density that allows them to exist without suffering a forced decay. Penergy too thin to spin into particles serves as the light density medium out of which particles can be created from a dense knot and into which they resolve upon decay.

Penergy is the source of physical mass, making $E=MC^2$ an accurate correlation between mass and Penergy. Despite our entire intergalactic medium being comprised of thin-density pure energy, that energy cannot be simply pulled out of the medium and used, except in the creation of particles. Consequently, the source of energy responsible for all Kinetic Energy *particle movements* is through Photon Energy and Field/ Potential Energy, both of which we characterize as Work Energy. We have mathematically found a way to measure Penergy and Work Energy in the same terms, but the two energies differ: Penergy is related to the content of particles, while Work Energy is related to the behavior of particles.

Penergy is, of course, much more than blackholes and particles. It holds many hidden dynamics and potentialities that only become apparent as blackholes and particles evolve. Without Penergy there would be no mass, fields, or forces, so energy in any form ultimately comes from Penergy. Penergy in all its forms has always defined our universe. When it was a speck, when it began spinning into SMBHs and particles, and when it smoothed out into galaxies of stars with the intergalactic medium in between, Penergy has always been the size, shape, substance, and definition of our universe. An understanding of Penergy certainly gives us a fresh perspective on the universe.

The CAGI at Work

This analysis allows us to peek inside the toolbox of the force driving cosmic evolution. Set against the backdrop of the Penergy medium, it appears that that the CAGI needs only three tools to carry out its work: 1) Tor particles — the substance from which to build complex matter; 2) fields — that which creates *forces* between particles that affects the interactions and relationships between them; and, 3) photons — giving movement energy to particles that provide them the means to couple/uncouple, form various phases (*i.e., gas, liquid, solid*), or transfer energy.

Like a fine painter meticulously applying color to his canvas, the CAGI molds matter into forms having surprising emergent qualities impossible to imagine given the apparent simplicity of the tools from which it must work. In much the way we appreciate a fine painter working with such simple tools, we are in awe of the CAGI's handiwork.

The CAGI Perspective

Those who theorize the ultimate destiny of the universe seem to consistently limit that destiny to three options: the universe will expand to a point and then collapse into a *big crunch*; the universe will expand to a point and then slow to a crawl, expanding gradually but never actually stopping; and finally, the universe will expand forever, until all matter is so separated that the universe becomes a cold, dark, lifeless, *abyss*.[2] Those theories are based solely on the physics relating to the geometry of space and the close balance between the strength of the energy expanding the universe in opposition to the strength of the gravitational energy shrinking the universe· Seldom do we see a theory that encompasses the richness of our cosmic evolution and how it may play a part in our cosmic future, influencing the ultimate destiny of our universe.

There are many ways to look at evolution in the broad sense, but the Combination and Growth Imperative (*the CAGI*) mentioned in Chapter Two is certainly an important perspective. Its basic tenets are simple and reasonable, making it difficult to ignore. Briefly, its premise is that the universe started with large swirls of Penergy that cascaded down to a band of single bits of left and right-spinning particles we have named Tors. Tors can create fields, a disturbance in the thin Penergy medium that encouraged Tors to *combine* nose to tail, manifesting the magnetic force that allowed the creation of Tor chains.

Opposite spinning Tors have a natural attraction for each other that allows them to *combine* with Tor chains, manifesting the electric force. This allowed Tryks to form and for Tryks to absorb and emit photon energy. The Tryks, utilizing the magnetic and electric forces, ultimately *combined* to produce the subatomic particles that eventually *combined* to form Atoms.

At each Level of matter the existing dominate form goes through its own evolutionary process, combining in different configurations which are being decimated by the elements within its environment. This process continues until a stable, strong, dominate form is produced that itself is capable of combining. During each Level's development, emergent qualities evolve that assist in that Level's survival, development, and ability to combine. Following the development of the atom, a variety of atoms *combined* to produce molecules, that *combined* to produce macromolecules, that *combined* to produce a cell.

To learn the details of the Combination and Growth Imperative, see *Journey of the Universe — A New Perspective on its Past, Present, and Future Evolution.*[3] It explains how cells combined into organisms and that at the top of the chain of organisms on earth is humankind. It then explains in detail how humankind might be the stable, dominate form of matter that ultimately combines into the Seventh Level of matter. It further explains in detail how that Seventh Level will ultimately combine into the Eighth Level. Much of this occurs while humans are living in space. It is an intriguing journey.

The reader may be having trouble imagining how some elements of humankind could possibly come together to create a larger form of matter. It is not as difficult as one might think. Looking over the history of humankind, especially within the last two-hundred years, the reader will understand how we are practically destined for that role and are currently very much on track to achieving it. Practically everything humankind has accomplished in its recent history has been unknowingly oriented toward humans combining into a larger entity. It is as if we are being guided by an unknown force, and we are globally doing exactly what we need to do to accomplish that combining process.

The *Journey* also looks out further into the future well past the Eighth Level and speculates on what cosmic evolution might be all about. The reader will find that there may be much more in store for humankind and the universe than the typically pronounced endings of a big crunch or a slow drift to an abyss. An understanding of the CAGI certainly gives us a new perspective on the universe.

The Tor Model Perspective

Cosmologist Carolyn Devereux tells us that the basic scientific process is drawing rational conclusions from observations in an objective way.[4] That has been the goal of this treatise. We started at the Big Bang with a blank slate, but rapidly filled it with observations and deductions, taking us quickly through the earliest phases of the universe. The journey required us to step off into the hazardous realm of creating new particles, which necessitated the creation of an entirely new model — the Tor Model. While developing the Tor Model we found it to be in significant contrast to many of the tenets of the Standard Model.

The Tor Model is not meant to undermine or take the place of any other model. It is food for thought in the development of ideas and theories regarding the unknowns and mysteries of the well-developed models we are currently using. Though much of the Tor Model makes sense, it does not yet have a mathematical basis, so the ideas expressed, though possibly valid, are likely to be slow to be accepted. The Tor Model nonetheless answers some of the important questions and issues plaguing both cosmology and particle physics. Here is a brief summary of some of those answers.

- **What were the initial conditions at the birth of the universe?**
  Beyond our belief the universe started from a very small speck of raw energy that for reasons unknown began to expand, there were no other initial conditions. Everything that subsequently happened, or was created, simply unfolds from changes in the density of the expanding raw energy, i.e., Penergy.

- **What is causing cosmic expansion *(The Dark Energy mystery)?***
  Penergy in a relatively light-density state naturally expands. As the early universe expanded the Penergy density thinned more and more until it is now our very thin interstellar medium, which is also the "dark energy" that has been expanding the universe since its inception.

- **How were supermassive blackholes (**SMBHs**) formed?**
  Penergy, like interstellar gas, with sufficient density naturally swirls, spins, and condenses. In the early universe when the Penergy was still very dense, clumps of it began swirling, ultimately creating a cascade of swirls. The larger of those swirls eventually spun themselves into pure-energy SMBHs.

- **How was the Cosmic Web formed?**
  The cascade of swirls created smaller and smaller swirls at their edges, ending in very tightly wound swirls. The large swirls became blackholes and the very small, tightly wound swirls became particles, both of which had the dynamics to sustain themselves. The intermediate sized swirls, however, could not sustain themselves in the very-thin density Penergy and eventually dissipated back into the Penergy medium. Their disappearance left large voids surrounded by various densities of SMBHs that became galaxies at the periphery of the voids, creating the cosmic web.

- **Why is the universe so homogenous *(The Horizon problem)?***
  The cascade of swirls and thinning Penergy eventually drew down to particle creation. The larger swirls were not quite developed into SMBHs, allowing the first particles to soar around the universe and begin combining. By the time the SMBHs were fully formed, the universe was homogeneous. The SMBH's then pulled in the existing matter and gases, creating the billions of galaxies we see today.

- **How were the first particles created, and why are there three generations?**
  At the end of the cascade of swirls of Penergy were very tiny swirls that spun into tightly wound particles. These tiny tornados are named Torons, or Tors for short. The first two generations of Tors and their composite particles were unable to survive in the thinning Penergy and dissipated back into the Penergy medium. They still come into existence momentarily when there is a knot of Penergy of sufficient density, but they decay quickly into lighter, stable particles. The third generation are the Tor composite particles that make up the quarks, electrons, neutrinos, and photons we are familiar with today.

- **What happened to the missing Anti-Matter?**
  The anti-matter isn't actually missing. Those particles are a part of an array of constituent elementary particles that make up the subatomic particles of today. Their presence is necessary for the creation of opposite charges *whose attractive/repulsive properties are necessary for the creation of both fields and particles.*

- **What is dark matter?**
  No one knows for sure. Perhaps it is particles left over from earlier evolutionary phases. Each of the first two generations and possibly a few generations before them,

went through their own evolutionary process of combining into different configurations until a strong, stable, combination was formed. The sea of remnants of those evolutionary trials were combinations that were strong enough to survive, but not adaptable enough to become a part of something larger. They would float around, bumping into each other until their charges were neutralized, rendering them "dark matter".

- **Why the disparity in the theorized level of vacuum energy and the actual measurement; a difference in the order of 10120 *(The Cosmological Constant problem)?***
  According to quantum physics, the vacuum of space is awash in virtual particles constantly popping in and out of existence and having a theorized vacuum energy density of 10^{105} *Joules per cubic centimeter. In the Tor Model, virtual particles only come into existence due to a fluctuation in the medium creating a small knot of Penergy, such as near blackholes and other large masses, or within the environment of a planet's atmosphere, such as our own. The measure of that limited energy density is likely much closer to the actual energy measured, which is only* 10^{-15} *Joules per cubic centimeter.*

- **What are emergent qualities?**
  Emergent qualities, as used in the Tor Model, are special attributes of matter that are unique to each evolutionary Level. They seem to evolve to assist in the development, survival, and combining capability of each new Level, though viewing the apparent simplicity of the Level of matter at the start of its evolutionary process, one would not expect such attributes to evolve.

The Tor-Model is rich in new ideas, many of which conflict with current theory. Perhaps some of those ideas will be the source of new theories that give both Standard

Models a better foundation. As noted by Astrophysicist Stuart Clark, when things have looked impossible before, the breakthrough has finally come when a scientist has thrown away a cherished assumption.[5] Perhaps throwing out the cherished assumption that our complex particles were created within the first second of the Big Bang will be the breakthrough the Standard Models needs to get on a better footing.

If Penergy is the stuff from which everything is made, and the CAGI is the story of how that Penergy has and will change, then the Tor Model is simply a theory of how those changes may have taken place in the early universe. An understanding of the Tor Model certainly gives us food for thought and a new perspective on the universe.

The Mathematical Perspective

Physicists view matter and energy and the relationships between their various manifestations as having a mathematical basis. It is believed or at least hoped that someday a single equation, or series of equations, will be found that simply and succinctly describe how everything is related — a theory of everything.

Expressing a theory of everything in a mathematical form will not be easy because the universe is constantly changing in ways we cannot anticipate. The natural core substance of the universe, Penergy, has not revealed all its secrets. Through the CAGI, evolutionary changes are taking place universe-wide and not all at the same time. With the evolution of each Level of matter, more emergent qualities will evolve having a significant impact on the appearance, capacity, and novel attributes of that Level and possibly future Levels. The presence of a new Level of matter affects the environment in which it resides, and the emergent qualities often have a substantial impact on determining what those effects will be. Consequently, emergent qualities make the future of matter and the environment in which it resides somewhat of an unknown.

Much of the exploration into how the universe works has been oriented toward a theory combining the four forces and peeling away matter to get to its core. This reductionist approach is fine for core particle discovery, but it will not by itself give us the picture we need for a theory of everything, unless we wish to confine the scope of that picture to solely the mechanics of the universe. Exposing the forces and building blocks, and their behavior and relationships, is only a small part of a bigger picture. The math of General Relativity and Quantum Mechanics is brilliant and accurate, but it only gives us measurements and expected experimental outcomes, not how the universe works in the big picture. Melding the two together won't likely solve that problem.

According to Physicist Michio Kaku, a theory of everything should take into consideration the initial conditions of the universe.[6] Theories as to how the universe began always start with several hand-picked conditions or premises upon which to build the theory. As we have seen, aside from the belief the universe started from a high-density speck of raw energy that began to expand, there is little need for any other initial conditions. The simpler we keep the initial conditions, the less problematic the consequent theories and explanations.

The problem with starting with any other initial conditions is that equations are then applied to describe those conditions and everything that follows must be consistent with those first equations. That creates a mathematical bias that limits our field of view and leads us down theoretical dead ends, such as particles being created before blackholes, which leaves unanswered questions, such as how were SMBHs created or how did the universe become so homogeneous? Many of the remaining similar mysteries of the universe can be traced back to a dependence on the prescribed initial conditions. Clearly, starting with few initial conditions beyond the initial speck of raw energy and its natural expansion is a better prospect for understanding the universe and avoiding blind alleys and unanswerable questions.

If a single equation is to describe everything, it would seem to necessitate on one side of the equation a symbol for Penergy, the substance from which everything is made. Perhaps it should also have a symbol for the CAGI — the ultimate description for how the universe has changed and will continue to change in the future. The symbol for the CAGI must include or course the concepts of *entropy* and *emergent properties*, as these are essential for understanding evolutionary change.

Mathematics is essential to our understanding, development, and existence. It is the best language for accurately describing the universe and is very important to us. Our endeavor to find the terms to describe the entire universe is certainly worthwhile. The sequence of the changes Penergy has gone through mentioned at the end of Chapter Seven may someday be reduced to a mathematical formulation, possibly representing at least a part of the theory of everything. An understanding of that new mathematical formulation, or an equally valid theory of everything, will certainly give us a broader perspective on the universe.

The Life Perspective

Humans naturally think of *life*, especially human life, as something special. We are conscious, intelligent beings, with the extraordinary capability of understanding the universe. The creation of life out of inanimate materials is nothing short of miraculous. We have a soul. We know God. Of course, we are special.

Perhaps we are. But perhaps we are simply complex organisms representing the earthbound Sixth Level of matter (*1st-Tors, 2nd-Tryks, 3rd-Sub A's, 4th-Atoms, 5th-Cells, 6th-Organisms*). We are endowed with consciousness and intelligence that no previous Level seems to possess. But these are simply emergent properties inherent in our Level of evolution. Every Level has its own emergent properties. The Seventh Level will be created when we humans combine (*if we survive without annihilating ourselves*). That Level will have its own emergent properties and likely feel themselves far superior, defining their existence in a term more appropriate to their attributes, and so endowed believe they are "special". It sounds a bit egotistical, but that is a good thing. The belief in their superiority will keep future Levels working to better their lot, keeping them motivated to develop the resources and means to create the next Level.

The special existence of humans carries over to the *anthropic principle* — the idea that life is special in the universe. The essential elements in the universe — matter, energy, forces, and their relationships — can all be measured and reduced to about thirty numbers called constants. Viewing those constants, it has been noted that they appear to be fine-tuned for the existence of life and that if we change one of the constants by a relatively small amount, we find that it makes the universe inhospitable to life.[7] Why is the specifications for the universe so hospital to life? This is known as the *fine-tuning problem.* It is one of nature's deepest mysteries.[8]

A belief the universe was designed to suit life and us humans seems compelling but is misplaced. Stephen Hawking recognized the argument, and after listing all the elements that had to be just right for us to exist, says, "Our universe and its laws appear to have a design that is both tailor-made and support us, and, if we are to exist, leaves little room for alteration."[9] Physicist Jim Baggott, however, reminds us we are very much a naturally evolved part of this reality, but we are not the reason for it.[10]

In the bigger picture, we humans are merely a product of the universe due to evolutionary natural selection. It's not that the universe created the conditions specifically for 'life' to exist, it's that the CAGI simply exploited those tenuous conditions and made life from them.

There are 100 billion galaxies, each with a 100-200 billion stars, each with an untold number of planets and moons. If even a small fraction of them is suitable for the biological experiment of life, it means the odds are we are not alone in the universe. If the CAGI theory is valid it would mean there are Level Fives, Sixes, and perhaps Sevens throughout the universe and we are definitely not alone. Alone or not, we are the trustees of the planet earth, and it's quite possible we are also a rare form of intelligent life. As a global society we need to acquire a new image of this bigger picture of who we are and what our responsibility is. We can no longer afford to act like school-yard children squabbling over possessions, territory, rights, rules, what God said, or whose God is supreme. Given the devastating technology we hold in our hands, that behavior is just not acceptable.

As a global society, we need to adopt a new perspective on *life*, what it means, and how we are going to respect it. If we are indeed special, we must realize the responsibility that comes with that privilege. Our primary goal must be to survive. That means we must take care of our planet, ourselves, and plan for the future. Assuring our survival as a species will mean someday leaving our planet and building colonies in space and on other planets or moons. This feat is best accomplished as a global effort. Creating such a global image, uniting in the common goal of manifesting our destiny in space, and working together to create a larger, more complex organism, will certainly give us a new perspective on the concept of *life*.

The Cosmic Perspective

It is apparent that a meaningful reality is more than simply understanding the mechanics of how matter and energy work on a micro or macro level. It is more than understanding the mechanics of how the universe has evolved, or how life has evolved. It is also apparent that reality depends on one's perspective and to understand the true reality of the universe — the cosmic reality — we must encompass as many perspectives as possible. Because the universe is forever changing in unanticipated ways due to the CAGI, a complete picture of a cosmic reality may have to encompass the universe from its birth to its final destiny. We may not know the full cosmic reality until we are much closer to the manifestation of that destiny.

It is also apparent that the destiny of the universe is far from it ending in either a big crunch or a cold abyss. While much of our cosmic history has been structured by the persistence of the CAGI and refined by natural selection, there are far too many wondrous things evolving for this complex cosmic dance to have been choreographed by mere chance. The creation of emergent qualities alone tells us there is much more to the capacity of the Penergy and the creativity of the CAGI than we can even imagine. The universe is surely on a journey. The most incredulous thing about it is that humankind may be in the envious position of being capable of figuring out its mysteries, as well as playing a small role in its destiny.

The persistence of the CAGI and its capacity for developing spectacular emergent qualities tells us there are fantastic experiences in our future. We need only to survive to become part of that cosmic destiny. To guide us in such an endeavor we need a picture of, or at least a vague image of, a cosmic reality. What early picture of cosmic reality can we derive from the lessons of the many perspectives we have examined so far?

A Cosmic Reality

The many perspectives we have examined so far tell us there is a much bigger picture to the universe than we have glimpsed thus far. The Tor Model has given us a picture of how the universe may have started and evolved, but more importantly it has given us an introduction to Penergy and the CAGI. Penergy is more than simply pure energy that creates and moves matter, it is the dynamic substance from which is molded the many unique qualities of our entire universe. Cosmologist Lawrence Krauss in summing dark energy says, 'It is natural to suspect that its nature is tied in some basic way to the origin of the universe. And all signs suggest that it will determine the future of the universe as well.'[11] Dr. Krauss's suspicions are right on. Dark matter (*Penergy*) has everything to do with the universe's unfolding, evolution, and destiny. For us, the prospects of witnessing more of Penergy's exciting creations and surprises are titillating.

The CAGI is the force that molds Penergy's unique qualities, bringing into existence fields, forces, connections, relationships, growth, intelligence, and life. Life is the Level of matter possessing the intelligence and technology to mold matter at an accelerated manner, transcending natural selection in its evolutionary process. Mathematics is the array of tools in our toolbox for material development and problem solving we will need to secure our survival.

From these perspectives it is obvious that we are on a wonderous journey. We are a life form possessing the intelligence and technology to influence our evolutionary development, allowing us to move off our planet to secure our future. In hindsight, our technological development has been unknowingly pushing us in that direction for the past two hundred years. That technology is allowing us to both modify ourselves in preparation for adapting to space and develop the physical means to live in space.

We may never know the exact history of how the early universe evolved, but while examining that realm we have learned a great deal about cosmic evolution and our own possible future. To realize that future we may have to shift the definition of reality away from the mechanical workings of the universe toward the bigger picture of the universe's changing evolution, more in line with a cosmic reality.

In the backdrop of the Penergy and the CAGI, we are products of an evolutionary journey that is inexorably building something very special. That something may very well be comprised of intelligence and technology, both of which we possess, creating the prospect of our playing a small part in that ultimate destiny. We know not the specifics of our immediate future, but our limited perspective suggests the overall course we must undertake to be a part of the cosmic destiny is for us to survive, to explore, and to evolve. The lessons we've learned from our different perspectives gives us the knowledge, blueprint, direction, and means to accomplish that goal. We are not only sightseers on this wondrous journey; we are a part of the journey itself.

* * *

Thank you for joining me on this extraordinary journey. It has been a wild ride and one that leaves us with a vision of the universe from a unique perspective. A perspective that only comes about if we are daring enough to view the universe not only from its mathematical description, but from an evolutionary basis derived from our observations and deductions. It is hoped that this new vision provides a better foundation not only for understanding the journey of the universe, but the journey of humankind, as well.

Acknowledgments

I would like to thank my faithful readers, Cary Nerelli, Steve Moore, Larry Firth, and Deborah Winter, my chief editor, Donna D. Lawson, and my illustrator, Anna Keene, for their time, good work, and encouragement.

Glossary

Accretion: The accumulation of particles into a massive object by gravitationally attracting matter, typically gaseous matter, in an accretion disk. Most astronomical objects, such as galaxies, stars, and planets, are formed by accretion processes.

Alpha Particle: A composite of two protons and two neutrons simultaneously ejected from a nucleus during nuclear decay.

Annihilate: A conversion of mass to pure energy when a particle of matter meets its anti-matter particle. The amount of energy involved is governed by Einstein's equation E=MC2.

Anthropic Principle: The idea we can make theories about the universe and the laws of physics based on the fact we exist in the universe.

Anti-matter: The twin of a matter particle with the same mass, but opposite spin/charge. When particles are created in pairs, one is matter and the other anti-matter, so all matter particles have an anti-matter twin. The two annihilate when they meet.

Baryon: Particles constructed of three quarks. Protons and Neutrons are baryons.

Beta Decay: The decay of a down quark into an up quark, electron, and neutrino.

Big Bang: The Standard Model's version of the origin of the universe from a small concentration of energy out of which space, time, and matter were abruptly created.

Blackhole: An object with a gravitational field so dense that not even light can escape it. Anything falling inside the black hole's outer boundary, called the *event horizon,* cannot escape. See also, *pure energy black hole.*

CAGI: The Combination and Growth Imperative that is driving the evolution of the universe.

Charge: See Electric charge.

Cosmic Microwave Background: The CMD is a relic of atom creation when photons were liberated. They are seen today as a low-level electromagnetic field pervading all space and measured to be 2.7 degrees Kelvin rather consistently.

Cosmology: The study of the origin, structure, and evolution of the universe.

Dark Energy: The energy theorized to be driving the expansion of the universe. There is no evidence for it other than it explains the expansion, but its source and nature are unexplained. The author believes dark energy is the thin-density, pure energy left over from the Big Bang after early blackhole and particle production consumed most of the denser pure energy.

Dark Matter: Unseen matter having no electromagnetic signature and therefore detectable only due to its gravitational influence. What it is made of is unknown. The author believes it may be made up of particle fragments that were left behind in the particle creation era after the Big Bang. For a full definition, see Chapter Two, *Dark Matter and Dynamic Gravity.*

Density: The ratio of a particular quantity, such as energy, to the volume in which it is contained.

Diffraction: The process by which waves are spread out as a result of passing through a narrow aperture or across an edge, typically accommodated by interference between the wave forms produced.

Dynamic Gravity: The author believes that gravity is generated directly by the spin of a black hole effecting the area outside and just inside the blackhole. Dynamic gravity effects the area deep inside the blackhole. It does not take effect until the black hole has matured, processed matter back to pure energy, and processed that pure energy back into a very dense pure energy. It is the very dense pure energy at its core that drives up the spin rate of the black hole that, in turn, shrinks the event horizon. For a full definition, see Chapter Two, *Dark Matter and Dynamic Gravity.*

Electric Charge: An attribute of electrons and quarks that gives them attractive and repulsive qualities. According to the Standard Model, the electric charge force is created by the respective particles exchanging virtual photons. According to the author's model, the electric charge force is created by a differential in the Tor particle count between the respective particles. For a full definition, see Chapter Four, *Electric Force.*

Electric Field: The field created by a charged particle or particles. For a full definition, see Chapter Five, *Particle Fields.*

Electromagnetic field: The interconnected electric and magnetic fields produced by an electric charge or photon in motion. For a full definition, see Chapter Five, *Particle Fields.*

Emergent Qualities: Properties and patterns not explainable by reducing a complex system to its simplest parts; behaviors that cannot be anticipated by viewing a system's sub-parts in isolation.

Energy: The ability to do work. For a full explanation, see Chapter Eight, *The Penergy Perspective.*

Entanglement: A quantum theoretical phenomena wherein a pair of particles interact and then once divided seem to have the capacity to communicate, even at large distances, without anything passing between them. For a full explanation, See Chapter Seven, *Particle Entanglement.*

eV: *see MeV.*

Event Horizon: The edge of a black hole that serves as the point at which if anything passes, it cannot escape the black hole's intense gravity.

Fermion: A generic term for particles having ½ spin, which includes quarks and leptons.

Field: An area of Penergy medium that is directly influenced by the vibrational presence of a particle or group of particles. For a full explanation, see Chapter Five, *Particle Fields.*

Galaxy: A loose gathering of stars and interstellar gases and particles held together by gravity.

Gamma Ray: A very high energy photon.

General Relativity: Einstein's theory of gravity, which cannot be distinguished from acceleration, causing a curvature of space-time, equating it with the geometry of space.

Gluon: A virtual particle that according to the Standard Model mediates the strong force holding quarks together.

Gravitational Field: The field created by the presence of mass. For a full explanation, see Chapter Five, *Particle Fields.*

Gravitational Wave: Oscillations in space-time caused by the movement of mass. Generally, they are undetected unless the movement is substantial, such as the collision of heavy objects such as black holes or neutron stars.

Graviton: The Standard Model's theoretical particle believed to be the messenger particle of a quantized gravitational field.

Hadron: The generic name for particles made up of quarks. Protons and Neutrons are Hadrons.

Heisenberg Uncertainty Principle: The theory that the more accurate we measure a particle's position, the less we know of its momentum, and vice versa. A similar relationship holds for energy and time, as well as various other observables.

Homogeneity Problem: Inability of the Standard Model's Big Bang theory (*without inflation*) to explain the present-day homogeneity of the universe. In the Big Bang theory, particles in the universe would not have had an opportunity to interact and reach thermal equilibrium. For a full definition, see Chapter Two, *Cosmic Homogeneity.*

Infinite: The notion something can go on forever, without limit or boundary.

Inflation: A theory of exponential expansion of the universe occurring within the first second after the Big Bang. It was developed in order to overcome the *homogeneity* problem.

Initial Mass: a measurement of the resistance to the acceleration of a body responding to a force.

Interference: Combination of two waves with the same frequency leading to reinforcement or diminution of joint intensity depending on the phase of the waves when meeting. Waves in phase meet constructively, and waves 180 degrees out of phase, destructively.

Interspatial Medium: The substance making up the space between stars and galaxies.

Kinetic Energy: Energy invested in the motion of a body.

Large Hadron Collider (LHC) is a particle collider located near Geneva, Switzerland, and is the largest of its kind, being a ring twenty-seven kilometers in circumference. It smashes particles together at high-speed allowing scientists to theorize from the collision debris how particles are constructed, relate, and behave.

Light Year: The distances traveled by light in one year: 9.46 x 10^{12} kilometers.

Magnetic Field: The field created by the alignment of charged particles, that in turn causes other charged particles to align. For a full definition, see Chapter Five, *Particle Fields.*

Meson: A composite particle made of a quark and an anti-quark.

MeV: Million electron Volts. A measurement used often by physicists to describe the mass or energy of a particle. An electron volt is the amount of energy acquired by one electron passing through an electric field having a one-volt differential. A MeV is 10^6 eV and A GeV is 10^9 eV.

Mirror-image: the view of an object as if a reflection in a mirror. That reflection would be opposite that of the object, but otherwise be the same in every visual respect.

Molecule: A stable combination of two or more atoms held together electromagnetically by the sharing of one or more electrons. A molecule has properties different from any of its atomic constituents.

Mutation: A mutation is an alteration in the nucleotide sequence of DNA, often caused by exposure to ultraviolet radiation.

Natural Selection: The process whereby organisms better adapted to their environment tend to survive and produce more offspring, perpetuating those survival traits. In the case of elementary particles, the best composite combinations tended to survive the frenzied environment and go on to become dominate and available to combine again.

Neutron Star: Late stage in the life of a star, reached when portions of the outer layers collapse under the influence of gravity. They form a dense inner core of neutrons created by protons and electrons forced together under extreme gravitational pressure.

Nucleus: The small, positively charged center of an atom comprised of protons and neutrons.

Pauli Principle. That principle encompasses the fact that no two like fermions (*electrons or quarks*) can occupy the same state, meaning carry on the same function, together. It is the reason that two electrons cannot occupy the same state in an atom. However, two electrons can appear in the same orbital ring if one is oriented with its spin pointing up, while the other is oriented with its spin pointing down.

Penergy: According to the author's model, Penergy is short for *pure energy* and is another name for raw energy, which is the energy from which everything in the universe is made.

Pion: a composite particle made up of two quarks believed to be a part of the force holding the nucleus of atoms together.

Polarization: The direction of which the amplitude of a wave is displaced.

Positron: The positive charged anti-particle to the electron.

Pure Energy Blackhole: The author's idea that the super-massive black holes at the heart of galaxies were spun out of pure energy. For a full definition, see Chapter One, *Alternative Theory for SMBH Creation.*

Quantum Field Theory (QFT): A theoretical framework that combines classical field theory, special relativity, and quantum mechanics. QFT treats particles as merely manifestations of their fields.

Quantum Mechanics: A mathematical formalism that gives a *probability* when calculating a particle's aspects.

Renormalization: a method used in quantum mechanics to remove unwanted infinities from equation solutions.

Singularity: A point in space-time when a physical quantity, such as mass or energy, becomes infinite.

Special Relativity: Einstein's theory relating space and time based on, 1) the laws of physics are invariant in all inertial frames of reference (*reference frames with no acceleration*), and 2) the speed of light in a vacuum is the same for all observers regardless of the motion of the light source or observer.

Spin: In general, and in the author's Tor Model, spin is the quality of something turning on its own axis. In particle physics it is a fundamental property of elementary particles that describes their state of rotation in terms of angular momentum times h/c^2. The h is Plank's constant: 6.626×10^{-34}, and c is the speed of light: 3.0×10^8 m/s.

Standard Model: Here used to represent the combination of the Standard Model of Particle Physics and the Cosmological Standard Model (*also known as the Concordance Model*) that together spell out the origin and history of particles and the universe.

Sub-A's: This is short for Subatomic Particles, referring to those Standard Model particles smaller than an atom.

Supermassive Blackhole (SMBH): Blackholes that are thousands, millions, and billions of times bigger than out sun. These huge blackholes are at the heart of every galaxy.

Supernova: The explosion of a dying star, spewing its heavy atoms out into space.

Supersymmetry: A theory of particle creation wherein all particles known today have a cousin particle much alike in appearance providing eloquent answers to particle issues, however, the cousin particles have never been seen and have not shown up in LHC experiments.

Symmetry: The ability of a system to appear unchanged despite undergoing some transformation.

Thermal Energy: is the flow or transference of energy between two bodies.

Thermal equilibrium: A state of uniform energy, or heat, throughout a given space, surface, or body.

Tor Model: The author's model of particle creation and the evolution of the early universe.

Tor: According to the author's Tor Model, a Tor is the first level of matter; the most elementary particle spun directly out of pure energy, and the particle from which all others are made.

Tryk: According to the author's Tor Model, a Tryk is the second level of matter, a combination of Tors and Tor-Chains, and the particles from which quarks and leptons are made.

Virtual Particle: A short lived particle created from a knot of energy occurring in the interspatial medium through natural fluctuations in the energy field. The knot of energy is insufficient to retain the particle in the current energy density.

Wave-Particle Duality: The theory that particles are both waves and point particles, displaying characteristics of both depending on how they are perceived or measured.

Notes and References

Introduction

1. Devereux, Carolyn, Cosmological Clues — Evidence for the Big Bang, Dark Matter, and Dark Energy, 2021, p.72.
2. Devereux, Carolyn, *Cosmological Clues — Evidence for the Big Bang, Dark Matter, and Dark Energy*, 2021, p.3.
3. Kaku, Michio, *The God Equation — The Quest for a Theory of Everything*, 2021, p.162.
4. Cliff, Harry, *How to Make an Apple Pie from Scratch*, 2021, p.302.
5. Stuart Clark, *The Unknown Universe — A New Exploration of Time, Space, and Modern Cosmology*, 2016, p.250.
6. Dine, Michael, *This Way to the Universe,* 2022, p.250-251
7. Randall, Lisa, *Knocking on Heaven's Door,* 2012, p.412
8. Deutsch, David, *The Beginning of Infinity*, 2011, p.vii.

Chapter One

1. Devereux, Carolyn, *Cosmological Clues — Evidence for the Big Bang, Dark Matter, and Dark Energy*, 2021, p.73. Dr. Devereux here is making the point the universe started from a singularity, which accounts for all the energy that exists today.
2. Cham, Jorge & Daniel Whiteson, *We Have No Idea,* 2018, p.32.
3. Ball, Philip, *Beyond Weird — Why Everything you Thought you knew about Quantum Physics is Different*, 2020, p.115. Dr. Ball more precisely says that in the earliest moments of the Big Bang the entire universe was smaller than an atom, which he characterizes as a quantum-mechanical entity.

4. Tyson, Neil deGrasse and Donals Goldsmith, *Origins — Fourteen Billion Years of Cosmic Evolution,* 2014, p.25.
5. Tyson, Neil deGrasse and Donals Goldsmith, *Origins — Fourteen Billion Years of Cosmic Evolution,* 2014, p.79. Dr. Tyson gives a good account of how after Einstein's theory of relativity was published, several scientists solved its equations and gave us the concepts of blackholes and singularities.
6. Perlov, Delia, and Alex Vilenkin, *Cosmology for the Curious,* 2017, p.94. The authors give a good account of Friedman's solutions to Einstein's equations and their predictions for an open and closed universe, and how those universes expand from a singularity.
7. Impey, Chris, *Einstein's Monsters — The Life and Times of Black Holes,* 2019, p.20.
8. Hawking, Stephen, *A Brief History of Time,* 1988, p.50. Dr. Hawking pointed out that the math was compelling and therefore the work became generally accepted and that nearly everyone assumed the universe started with a Big Bang singularity. He went on to say that it was ironic that he changed his mind and was trying to convince other physicists that there was in fact no singularity at the beginning of the universe.
9. Devereux, Carolyn, *Cosmological Clues — Evidence for the Big Bang, Dark Matter, and Dark Energy,* 2021, p.73.
10. James, Tim, *Fundamental — How Quantum and Particle Physics Explain Absolutely Everything,* 2020, p.205. Dr. James explains that infinity is not a real thing in physics. "It exists in abstract mathematics, but in the universe, there are no infinities".
11. Impey, Chris, *Einstein's Monsters — The Life and Times of Black Holes,* 2019, p.19.
12. Kaku, Michio, *The God Equation — The Quest for a Theory of Everything,* 2021, p.57.
13. Impey, Chris, *Einstein's Monsters — The Life and Times of Black Holes,* 2019, p.14.
14. Beckman, Milo, *Math Without Numbers,* 2021, p.99.

15. Kaku, Michio, *The God Equation, The Quest for a Theory of Everything*, 2021, p.117.
16. Devereux, Carolyn, *Cosmological Clues*, 2021, p.3. Dr. Devereux says that the Big Bang contained all the energy of the universe, and that within the first millionth of a second, all the basic elements and forces that we now see in the universe were formed producing our subatomic particles.
17. Seife, Charles, *Alpha & Omega — The Search for the Beginning and End of the Universe,* 2003, P.216.
18. Devereux, Carolyn, Cosmological Clues, 2021, p.65. Dr. Devereux admits scientists do not know what dark energy is, or how it can even exist, but she is confident it does exist.
19. Devereux, Carolyn, *Cosmological Clues — Evidence for the Big Bang, Dark Matter, and Dark Energy*, 2021, p.73.
20. Tucker, Wallace H., *Chandra's Cosmos*, 2017, p.26.
21. Randall, Lisa, *Knocking on Heaven's Door — How Physics and Scientific Thinking Illuminate the Universe and the Modern World,* 2012, P.99.
22. Livio, Mario, *The Accelerating Universe — Infinite Expansion, the Cosmological Constant, and the Beauty of the Cosmos*, 2000, p.126.
23. Impey, Chris, *How it Began — A Time Traveler's Guide to the Universe*, 2012, p.333
24. Baggott, Jim, *Origins — The Scientific Story of Creation*, 2015, p.20. Dr. Baggott relays this fact incidental to discussing an experiment (*the Gravity Probe*) developed to prove Einstein's theory of General Relativity by proving the earth is bending the spacetime around it.
25. Cham, Jorge & Daniel Whiteson, *We Have No Idea,* 2018, p.91. The authors describe how the LIGO (Laser Interferometer Gravitational-Wave Observatory) works, and how its success beautifully confirmed Einstein's picture of gravity bending space.
26. Cham, Jorge & Daniel Whiteson, *We Have No Idea*, 2018, p.90.

27. Cham, Jorge & Daniel Whiteson, *We Have No Idea,* 2018, p.103. The authors make a good argument for what they are calling *space goo* as being the something that exists in the vacuum of space, and the something that gives space its physical nature and the ability to ripple, bend, wave, and expand.
28. Sutter, Paul S., *Your Place in the Universe — Understanding Our Big Messy Existence,* 2018, p.86.
29. Tucker, Wallace H., *Chandra's Cosmos,* 2017, p.57. Dr. Tucker tells us that Dark Energy has the peculiar property of having negative energy. He points out that normally when a gas expands, its energy decreases. Dark energy does just the opposite: the amount of dark energy keeps increasing as the universe expands.
30. Randall, Lisa, *Dark Matter and the Dinosaurs — The Astounding Interconnectedness of the Universe,* 2015, p.39. Speaking of the Hubble Constant, Dr. Randall says, 'It is a constant in the sense that today, its value in space is everywhere the same. But actually the Hubble parameter is not constant. It changes with time. Earlier in the universe, when things were denser and gravitational effects were stronger, the universe expanded far more rapidly than it does today.'
31. Impey, Chris, *Einstein's Monsters — The Life and Times of Black Holes,* 2019, p.150.
32. Impey, Chris, *Einstein's Monsters — The Life and Times of Black Holes,* 2019, p.133-134.
33. Impey, Chris, *Einstein's Monsters — The Life and Times of Black Holes,* 2019, p.54.
34. Impey, Chris, *Einstein's Monsters — The Life and Times of Black Holes,* 2019, p.221.
35. Impey, Chris, *Einstein's Monsters — The Life and Times of Black Holes,* 2019, p.53. Dr. Impey says that the maximum rate at which matter can be added to a blackhole is rather low; in a year it can grow no more than a third of the moon's mass. At that rate it would take 30 million years to double in mass. Dr. Impey points out on

page 146 that supermassive blackholes found to have been formed within the first billion years after the Big Bang are at odds with maximum growth rate a blackhole is determined to grow.

36. Perlov, Delia, and Alex Vilenkin, *Cosmology for the Curious*, 2017, p.181.
37. Impey, Chris, *Einstein's Monsters — The Life and Times of Black Holes*, 2019, p.146.
38. Impey, Chris, *Einstein's Monsters — The Life and Times of Black Holes*, 2019, p.103.
39. Impey, Chris, *Einstein's Monsters — The Life and Times of Black Holes*, 2019, p.146.
40. Impey, Chris, *How it Began — A Time Traveler's Guide to the Universe*, 2012, p.165. Dr. Impey says that computer simulations show that the merger process scrambles the near circular orbits of the stars in the two [spiral] disks and scatters them into a near spherical cloud.
41. Impey, Chris, *How it Began — A Time Traveler's Guide to the Universe*, 2012, p.152.
42. Impey, Chris, *How it Began — A Time Traveler's Guide to the Universe*, 2012, p.165. Dr. Impey says that scientists wondered what caused galaxies to be elliptical in shape, and the verdict finally came down on the side of mergers [probably through computer simulations].
43. Greene, Brian, *Until the End of Time*, 2020, p.287.
44. Impey, Chris, *How it Began — A Time Traveler's Guide to the Universe*, 2012, p.206.
45. Tucker, Wallace H., *Chandra's Cosmos*, 2017, p.124.
46. Impey, Chris, *How it Began — A Time Traveler's Guide to the Universe*, 2012, p.154-156. Chapter Seven of the book is entitled *Cosmic Architecture*, and it goes into how the galaxies are shaped and the possible causes for their distribution. The description I use is drawn from several pages in the book, starting at page 154.
47. Barnes, Luke A. & Geraint F. Lewis, *The Cosmic Revolutionary's Handbook — Or: How to Beat the Big Bang*, 2020, p.238.

48. Tucker, Wallace H., *Chandra's Cosmos*, 2017, p.123.
49. Sutter, Paul S., *Your Place in the Universe — Understanding Our Big Messy Existence*, 2018, p.162
50. Sutter, Paul S., *Your Place in the Universe — Understanding Our Big Messy Existence*, 2018, p.165.
51. Perlov, Delia, and Alex Vilenkin, *Cosmology for the Curious*, 2017, p.195.
52. Impey, Chris, *Einstein's Monsters — The Life and Times of Black Holes*, 2019, p.143.
53. Impey, Chris, *Einstein's Monsters — The Life and Times of Black Holes*, 2019, p.233.
54. Tucker, Wallace H., *Chandra's Cosmos*, 2017, p.61.
55. Devereux, Carolyn, *Cosmological Clues*, 2021, p.62. According to Dr. Devereux, the voids in the cosmic web are about 100 million light years across and they make up about 90% of the universe. Pages 61-62 give a good description of the cosmic web.
56. Gott, J. Richard, *The Cosmic Web*, 2016, p.104
57. Seife, Charles, *Alpha & Omega — The Search for the Beginning and End of the Universe*, 2003, P.105.
58. Impey, Chris, *Einstein's Monsters — The Life and Times of Black Holes*, 2019, p.239.
59. Livio, Mario, *The Accelerating Universe — Infinite Expansion, the Cosmological Constant, and the Beauty of the Cosmos*, 2000, p.66.
60. Randall, Lisa, *Knocking on Heaven's Door — How Physics and Scientific Thinking Illuminate the Universe and the Modern World*, 2012, P.123.

Chapter Two

1. Devereux, Carolyn, *Cosmological Clues — Evidence for the Big Bang, Dark Matter, and Dark Energy*, 2021, p.116.
2. Devereux, Carolyn, *Cosmological Clues — Evidence for the Big Bang, Dark Matter, and Dark Energy*, 2021, p.86.
3. Devereux, Carolyn, *Cosmological Clues — Evidence for the Big Bang, Dark Matter, and Dark Energy*, 2021, p.6.

4. Devereux, Carolyn, *Cosmological Clues — Evidence for the Big Bang, Dark Matter, and Dark Energy,* 2021, p.112.
5. Devereux, Carolyn, Cosmological Clues — Evidence for the Big Bang, Dark Matter, and Dark Energy, 2021, p.74.
6. Greenstein, George, *Symbiotic Universe — Life and Mind in the Cosmos,* 1988, p.164.
7. Hawking, Stephen & Leonard Mlodinow, *The Grand Design,* 2010, p.129.
8. Guth, Alan H., *The Inflationary Universe — The Quest for a New Theory of Cosmic Origins*, p.238
9. James, Tim, Astronomical -From Quarks to Quasars, the Science of Space at Its Strangest, 2020, p.57.
10. Carroll, Sean, *The Big Picture — The Origins of Life, Meaning, and the Universe Itself,* 2017, p.308.
11. Tyson, Neil deGrasse and Donals Goldsmith, *Origins — Fourteen Billion Years of Cosmic Evolution,* 2014, p.138.
12. Livio, Mario, *The Accelerating Universe — Infinite Expansion, the Cosmological Constant, and the Beauty of the Cosmos,* 2000, p.67.
13. Tyson, Neil deGrasse and Donals Goldsmith, *Origins — Fourteen Billion Years of Cosmic Evolution,* 2014, p.26.
14. Kaku, Michio, *Physics of the Impossible — A Scientific Exploration into the World of Phasers, Force Fields, Teleportation, and Time Travel,* 2008, p.247. In his discussion as to whether our universe could have been created from nothingness, Dr. Kaku says that many physicists have pointed out that it is astonishing that the total amount of positive charges and negative charges in the universe comes out to be exactly zero, at least to within experimental accuracy.
15. Butterworth, Jon, *Atom Land — A Guided Tour Through the Strange (and Impossibly Small) World of Particle Physics,* 2019, p.193.
16. Butterworth, Jon, *Atom Land — A Guided Tour Through the Strange (and Impossibly Small) World of Particle Physics,* 2019, p.275. Dr. Butterworth speculates on

the prospect of the quarks, leptons, and bosons of the Standard Model may contain smaller constituents, just as the atoms on the Periodic Table turned out to be made of other things.

17. Ford, Kenneth W., *The Quantum World — Quantum Physics for Everyone,* 2004, p.61.
18. Munowitz, Michael, *Knowing — The Nature of Physical Law,* 2005, p.34.
19. Cham, Jorge & Daniel Whiteson, *We Have No Idea,* 2018, p.212.
20. Close, Frank, *Antimatter,* 2018, p.84
21. Kaku, Michio, *Physics of the Impossible — A Scientific Exploration into the World of Phasers, Force Fields, Teleportation, and Time Travel,* 2008, p.182. In his discussion on the prospects of building an antimatter rocket, Dr. Kaku describes the construction of a 'Penning Trap' containing liquid helium and nitrogen that would store a trillion anti-protons. He says parenthetically, 'At very low temperatures, the wavelength of the anti-protons is several times longer than the wavelength of the particles in the container walls, so the anti-protons would mainly reflect off the walls without annihilating themselves.'
22. Davies, Paul, *The Cosmic Blueprint — New Discoveries in Nature's Creative Ability to Order the Universe,* 1989, p.5.
23. Alexander, Stephon, *Fear of a Black Universe — An Outsiders Guide to the Future of Physics,* 2021, p.65.
24. Schumm, Bruce A., *Deep Down Things — The Breath-Taking Beauty of Particle Physics,* 2004, p.88.

Chapter Three

1. Cham, Jorge & Daniel Whiteson, *We Have No Idea — A Guide to the Unknown Universe,* 2018, p.212.
2. Fritzsch, Harald, *The Fundamental Constants — A Mystery of Physics,* 2005, p.27.

3. Butterworth, Jon, *Atom Land — A Guided Tour Through the Strange (and Impossibly Small) World of Particle Physics,* 2019, p.120.
4. Hawking, Stephen, *A Brief History of Time,* 1988, p.167.
5. Cham, Jorge & Daniel Whiteson, *We Have No Idea — A Guide to the Unknown Universe,* 2018, p.50.
6. Dine, Michael, *This Way to the Universe,* 2022, p.238.
7. Cham, Jorge & Daniel Whiteson, *We Have No Idea — A Guide to the Unknown Universe,* 2018, p.54.
8. Wilczek, Frank, *Fundamentals — Ten Keys to Reality,* 2021, p.57. Dr. Wilczek tells us that, 'The Higgs particles, whose discovery was a major triumph for twenty-first century physics, is highly unstable. It lives for only about 10^{-22} second. Thus, in order to discern evidence for its existence, physicists had to reconstruct events on that time scale.'
9. Dine, Michael, *This Way to the Universe,* 2022, p.138.
10. Krauss, Lawrence, *The Greatest Story Ever Told — So Far — Why are We Here,* 2018, p.271.
11. Kaku, Michio, *The God Equation — The Quest for a Theory of Everything,* 2021, p.89.
12. Hossenfelder, Sabine, *Lost in Math — How Beauty Leads Physics Astray,* 2020, p.37-38. Dr. Hossenfelder says, '... quantum fluctuations make a huge contribution to the Higgs mass. Contributions like this are normally small, but for the Higgs they lead to a mass much larger than what is observed — too large, indeed, by a factor of 10^{14}. Not a little bit off, but dramatically, inadmissibly wrong.' She goes on to explain how the numbers can be adjusted and points out that the final number requires an explanation, and that numbers requiring an explanation are called *fine-tuned.* She says, 'In the Standard Model, the Higgs mass is not natural, which makes it ugly.'
13. Krauss, Lawrence, *The Greatest Story Ever Told — So Far — Why are We Here,* 2018, p.255-256. Dr. Krauss tells us, 'These particles that interact with the Higgs

background field then experience a kind of resistance to their motion that slows their travel to less than the speed of light-just as a swimmer in molasses will move more slowly than a swimmer in water. Once they are moving at sub-light speed, the particles behave as if they are massive.'

14. Krauss, Lawrence, *The Greatest Story Ever Told — So Far — Why are We Here,* 2018, p.256. Here Dr. Krauss is pointing out that 'For every new field in nature, at least one new type of elementary particle must exist with that field.' He describes this as a central property of quantum field theory.
15. Fritzsch, Harald, *The Fundamental Constants — A Mystery of Physics,* 2005, p.113.
16. Lindley, David, *The Dream Universe — How Fundamental Physics Lost its Way,* Doubleday, New York, 2020, p.151.
17. Impey, Chris, *Einstein's Monsters — The Life and Times of Blackholes,* 2019, p.174. Dr. Impey tells us, '...in Einstein's theory [of General Relativity], mass couple to the geometry of space-time. In 1918 in was predicted that the rotation of a massive object would distort space-time... This twisting of the contours of space is called frame dragging.' He goes on to relate the story of the experiment than confirmed this prediction, and how Einstein's prediction of space-time curvature and frame dragging was confirmed.
18. Wilczek, Frank, *Fundamentals — Ten Keys to Reality,* 2021, p.74.

Chapter Four

1. Fritzsch, Harald, *The Fundamental Constants — A Mystery of Physics,* 2005, p.116.
2. Barrow, John D. and Joseph Silk, *The Left Hand of Creation — The Origin and Evolution of the Expanding Universe,* 1993, p.218.
3. Dine, Michael, *This Way to the Universe,* 2022, p.28.

4. Huang, Kerson, *Fundamental Forces of Nature — The Story of Gauge Fields*, 2007, p.99.
5. Hawking, Stephen, A Brief History of Time, 1988, p.68-69.
6. Baggott, Jim, *Origins — The Scientific Story of Creation*, 2015, p.79-80.
7. Butterworth, Jon, *Atom Land — A Guided Tour Through the Strange (and Impossibly Small) World of Particle Physics*, 2019, p.217.
8. Hawking, Stephen, *A Brief History of Time*, 1988, p.157.
9. Devereux, Carolyn, *Cosmological Clues — Evidence for the Big Bang, Dark Matter, and Dark Energy*, 2021, p.93.
10. Kaku, Michio, *The God Equation — The Quest for a Theory of Everything*, 2021, p.104. Dr. Kaku tells us that after scientists introduced the theory of the graviton that, 'Sadly, the bag of tricks painfully accumulated by physicists for the past seventy years to eliminate these infinities failed for the graviton. Here, physicists hit a brick wall.'
11. Ford, Kenneth W., *The Quantum World — Quantum Physics for Everyone*, 2005, p.86-87. In his discussion on bosons and the creation/annihilation of particles, he says, '...the interaction event is a truly catastrophic event in which every particle is either annihilated or created.' In referring to a Feynman diagram he says, 'At point A, the incoming electron is destroyed, a photon is created, and a *new* electron is created.' What he is saying overall is that every particle interaction creates of a virtual boson and two new particles. This phenomenon was also noted in *We Have No* Idea by Jorge Cham and Daniel Whiteson, who say on page 212, 'That means, in a way, all particle interactions result in annihilation of the original particles into new particles.
12. Musser, George, *Spooky Action at a Distance — The Phenomenon that Reimagines Space and Time and What it Means for Black Holes, the Big Bang, and Theories of Everything*, 2015, p.73. Dr. Musser does not make this observation incidental to an argument for force being

conveyed by fields rather than virtual particles. Nonetheless, if his observation is accurate, it could be used, as I am using it, to suggest that force showing a slight delay when conveyed, better suits the idea of it being conveyed through a medium such as Penergy, rather than by virtual boson exchange.

13. Krauss, Lawrence, *The Greatest Story Ever Told — So Far — Why are We Here,* 2018, p.104. Dr. Krauss here is making the point that since a virtual particle's time in existence is inversely proportional to its energy, a photon carrying very little energy could travel as far as Alpha Centauri. My point is that given that is true, the electron on earth could also have the same relationship with every electron between earth and Alpha Centauri, making the constant exchange of photons ridiculous, or if a constant exchange is not taking place with all those electrons, there is nothing more than an intermittent force between them.
14. James, Tim, *Fundamental — How Quantum and Particle Physics Explain Absolutely Everything,* 2020, p.180-181.
15. Kaku, Michio, *The God Equation — The Quest for a Theory of Everything,* 2021, p.138.
16. Perlov, Delia, and Alex Vilenkin, *Cosmology for the Curious,* 2017, p.306.
17. Cham, Jorge & Daniel Whiteson, *We Have No Idea — A Guide to the Unknown Universe,* 2018, p.222.
18. Butterworth, Jon, *Atom Land — A Guided Tour Through the Strange (and Impossibly Small) World of Particle Physics,* 2019, p.137.
19. Fritzsch, Harald, *The Fundamental Constants — A Mystery of Physics,* 2005, p.167.
20. Wilczek, Frank, *Fundamentals — Ten Keys to Reality,* 2021, p.89.
21. Alexander, Stephon, *Fear of a Black Universe — An Outsiders Guide to the Future of Physics,* 2021, p.70
22. Perlov, Delia, and Alex Vilenkin, *Cosmology for the Curious,* 2017, p.202.

Chapter Five

1. Krauss, Lawrence, *The Greatest Story Ever Told — So Far — Why are We Here,* 2018, p.109.
2. Cham, Jorge & Daniel Whiteson, *We Have No Idea,* 2018, p.65.
3. Cham, Jorge & Daniel Whiteson, *We Have No Idea, 2018,* p.113.
4. Carroll, Sean, *The Big Picture — The Origins of Life, Meaning, and the Universe Itself,* 2017, p.173. Dr. Carroll tells us that, '...a field is something that stretches all throughout space, taking on some particular value at every point. Modern physics says that the particles and the forces that make up all atoms all arise out of fields.'
5. Munowitz, Michael, *Knowing — The Nature of Physical Law,* 2005, p.38.
6. Musser, George, *Spooky Action at a Distance — The Phenomenon that Reimagines Space and Time and What it Means for Black Holes, the Big Bang, and Theories of Everything,* 2015, p.134. Dr. Musser in his discussion of the development of quantum field theory, says that physicists combined elements of both quantum mechanics and relativity theory, performing a shotgun marriage with unforeseeable consequences. 'To this day, physicists struggle to fathom what quantum field theory is telling them about the world.' In the same paragraph he goes on to say with regard to quantum field theory, 'It has a deserved reputation as the most mathematically badass subject in the sciences. Even experts hang on for dear life.'
7. Munowitz, Michael, *Knowing — The Nature of Physical Law,* 2005, p.322.
8. Carroll, Sean, *The Big Picture — The Origins of Life, Meaning, and the Universe Itself,* 2017, p.176
9. Smith, Timothy Paul, *Hidden Worlds — Hunting for Quarks in Ordinary Matter,* 2003, p.13.
10. Kaku, Michio, *The God Equation, The Quest for a Theory of Everything,* 2021, p.58. Dr. Kaku in his relating the

history of discovery in particle physics from the perspective of what scientists believed to be true at the time, says, 'Light was made of photons, which are quanta, or particles, but each photon created fields around it (the electric and magnetic fields). These fields, in turn, were shaped like waves and obeyed Maxwell's equations. We now have a beautiful relationship between particles and the fields that surround it.' Although he was not arguing specifically that it is the fields that are doing the waving, his choice of words certainly suggests that is the case.

11. Ball, Philip, Beyond Weird — Why Everything You Thought About Quantum Physics is Different, 2018, p.169.
12. Becker, Adam, What is Real — The unfinished quest for the meaning of quantum physics,2019, p.145.
13. Ball, Philip, Beyond Weird — Why Everything You Thought About Quantum Physics is Different, 2018, p.165.
14. Ball, Philip, Beyond Weird — Why Everything You Thought About Quantum Physics is Different, 2018, p.64.
15. McTaggart, Lynne, The Field — The Quest for the Secret Force of the Universe, 2002, p.43.
16. Huang, Kerson, Fundamental Forces of Nature — The Story of Gauge Fields, 2007, p.96.
17. Baggott, Jim, Origins — The Scientific Story of Creation, 2015, p.45.
18. Davies, Paul and John Gribbin, The Matter Myth — Dramatic Discoveries that Challenge our Understanding of Physical Reality, 1992, p.176.
19. Alexander, Stephon, Fear of a Black Universe — An Outsiders Guide to the Future of Physics,2021, p.113.
20. Cham, Jorge & Daniel Whiteson, We Have No Idea, 2018, p.37. In their discussion of the picture made by the Cosmic Microwave Background (CMB), the authors say the picture is very sensitive to the proportion of

dark matter, dark energy, and regular matter, and if you change the proportions the picture comes out differently. Based on the current picture scientists are seeing in the CMB, one would need about 5 percent regular matter, 27 percent dark matter, and 68 percent dark energy. 'Anything else would give a different picture than what we observe.'

21. Cham, Jorge & Daniel Whiteson, We Have No Idea, 2018, p.266.
22. Quinn, Helen R. & Yossi Nir, The Mystery of the Missing Anti-matte, 2008, p.27.
23. Han, M Y, Quarks and Gluons — A Century of Particle Charges, 1999, p.70.
24. www.brittanica.com/science/beta-decay.
25. Ford, Kenneth W., The Quantum World — Quantum Physics for Everyone, 2005, p.86.
26. Munowitz, Michael, Knowing — The Nature of Physical Law, 2005, p.32
27. Krauss, Lawrence, The Greatest Story Ever Told — So Far — Why are We Here, 2018, p.81
28. Cowen, Ron, Gravity's Century — From Einstein's Eclipse to Images of Black Holes, p.93
29. Clegg, Brian, Gravity, 2016, p.128
30. Munowitz, Michael, Knowing — The Nature of Physical Law, 2005, p.106
31. Geach, James, Five Photons — Remarkable Journeys of Light Across Space and Time, p.113.
32. Clegg, Brian, Gravity, 2016, p.174
33. Close, Frank, Antimatter, 2018, p.28.

Chapter Six

1. Beckman, Milo, Math Without Numbers, 2021, p.191.
2. Kaku, Michio, *The God Equation, The Quest for a Theory of Everything*, 2021, p.101.
3. Cham, Jorge & Daniel Whiteson, *We Have No Idea*, 2018, p.19-20.

4. Butterworth, Jon, *Atom Land — A Guided Tour Through the Strange (and Impossibly Small) World of Particle Physics*, 2019, p.106.
5. Impey, Chris, *Einstein's Monsters — The Life and Times of Black Holes,* 2019, p.174. (See reference note #8.)
6. Cham, Jorge & Daniel Whiteson, *We Have No Idea,* 2018, p.73-74.
7. Impey, Chris, *Einstein's Monsters — The Life and Times of Black Holes,* 2019, p.174. Dr. Impey tells says, 'Gravity does not depend on rotation in Newton's theory. But in Einstein's theory, mass couples to space-time. In 1918 it was predicted that the rotation of a massive object would distort space-time, making the orbit of a smaller nearby object precess, like the pivoting of a spinning top. This twisting of the contours of space is called frame-dragging.'
8. Hawking, Stephen, Black Holes and Baby Universes, and Other Essays, 1994, P.75
9. Nomura, Yasunori & Bill Poirier & John Terning, *Quantum Physics, Mini Black Holes, and the Multiverse — Debunking Common Misconceptions in Theoretical Physics*, 2018, p.97
10. Nomura, Yasunori & Bill Poirier & John Terning, *Quantum Physics, Mini Black Holes, and the Multiverse — Debunking Common Misconceptions in Theoretical Physics*, 2018, p.98
11. Impey, Chris, *Einstein's Monsters — The Life and Times of Black Holes,* 2019, p.176.
12. Devereux, Carolyn, *Cosmological Clues*, 2021, p.91. Dr. Devereux in her discussion on the topic of MOND, an alternative accounting for galaxy rotation without dark matter, she says, 'Mond not only explains the rotation curves without the need for dark matter. It does more. From the equations it can be shown that the mass of a galaxy is related to its velocity of rotation (v) to the power (v^4).' This observation is in keeping with the Tor Model's theory of blackhole mass being related to the blackholes spin rate.

13. Timothy Paul Smith's, *Hidden Worlds — Hunting for Quarks in Ordinary Matter*
14. Kisslinger, Leonard S., *Astrophysics and The Evolution of the Universe*, 2017, p.38-39.
15. Wilczek, Frank, *A Beautiful Question — Finding Nature's Deep Design*, 2015, p.260.
16. Kaku, Michio, *The God Equation, The Quest for a Theory of Everything*, 2021, p.92.
17. Krauss, Lawrence, *The Greatest Story Ever Told — So Far — Why are We Here,* 2018, p.253.
18. Perlov, Delia, and Alex Vilenkin, *Cosmology for the Curious*, 2017, p.190.
19. Baggott, Jim, *Origins — The Scientific Story of Creation,* 2015, p.66.
20. Tucker, Wallace H., *Chandra's Cosmos*, 2017, p.21.
21. Baggott, Jim, *Origins — The Scientific Story of Creation,* 2015, p.64.
22. Rothman, Tony, *A Little Book About the Big Bang*, 2022, p.73.
23. Cham, Jorge & Daniel Whiteson, *We Have No Idea,* 2018, p.238.
24. Gott, J. Richard, *The Cosmic Web*, 2016, p.25-26.
25. Perlov, Delia, and Alex Vilenkin, *Cosmology for the Curious*, 2017. P.232

Chapter Seven

1. Smolin, Lee, The Trouble with Physics — The Rise of String Theory, the Fall of Science, and What Comes Next, 2007, p.7.
2. Wilczek, Frank, *A Beautiful Question — Finding Nature's Deep Design*, 2015, p.9.
3. Becker, Adam, *What is Real — The unfinished quest for the meaning of quantum physics,* 2019, p.49.
4. Clegg, Brian, *The God Effect — Quantum Entanglement, Science's Strangest Phenomenon,* 2006, p.220.
5. Kaku, Michio, *Visions- How Science will Revolutionize the 21st Century,* 1997, p.109.

6. Randall, Lisa, *Knocking on Heavens Door — How Physics and Scientific Thinking Illuminate the Universe and the Modern World,* 2012, p.23.
7. Becker, Adam, *What is Real — The unfinished quest for the meaning of quantum physics,* 2019, p.14.
8. Smolin, Lee, *Einstein's Unfinished Revolution — The Search for What Lies Beyond the Quantum,* 2019, p.145.
9. Krauss, Lawrence, *The Greatest Story Ever Told — So Far — Why are We Here,* 2018, p.90.
10. Smolin, Lee, *Einstein's Unfinished Revolution — The Search for What Lies Beyond the Quantum,* 2019, p.21.
11. Chaisson, Eric J., *Cosmic Evolution — The Rise of Complexity in Nature,* 2001, p.34.
12. Nomura, Yasunori & Bill Poirier & John Terning, *Quantum Physics, Mini Black Holes, and the Multiverse — Debunking Common Misconceptions in Theoretical Physics,* 2018, p.21.
13. Krauss, Lawrence, *The Greatest Story Ever Told — So Far — Why are We Here,* 2018, p.91.
14. Nomura, Yasunori & Bill Poirier & John Terning, *Quantum Physics, Mini Black Holes, and the Multiverse — Debunking Common Misconceptions in Theoretical Physics,* 2018, p.21.
15. Nomura, Yasunori & Bill Poirier & John Terning, *Quantum Physics, Mini Black Holes, and the Multiverse — Debunking Common Misconceptions in Theoretical Physics,* 2018, p.21.
16. Fritzsch, Harald, *The Fundamental Constants — A Mystery of Physics,* 2005, p.26.
17. Huang, Kerson, *Fundamental Forces of Nature — The Story of Gauge Fields,* 2007, p.97.
18. Carroll, Sean, *The Big Picture — The Origins of Life, Meaning, and the Universe Itself,* 2017, p.177.
19. Ford, Kenneth W., *The Quantum World — Quantum Physics for Everyone,* 2005, p.246.
20. James, Tim, *Fundamental — How Quantum and Particle Physics Explain Absolutely Everything,* 2020, p.50.

21. Nomura, Yasunori & Bill Poirier & John Terning, *Quantum Physics, Mini Black Holes, and the Multiverse — Debunking Common Misconceptions in Theoretical Physics,* 2018, p.16.
22. Geach, James, *Five Photons — Remarkable Journeys of Light Across Space and Time,* 2018, p.73.
23. James, Tim, *Fundamental — How Quantum and Particle Physics Explain Absolutely Everything*, 2020, p.52.
24. Ford, Kenneth W., *The Quantum World — Quantum Physics for Everyone,* 2005, p.185.
25. Ball, Philip, *Beyond Weird — Why Everything you Thought you knew about Quantum Physics is Different,* 2020, p.39.
26. Ball, Philip, *Beyond Weird — Why Everything you Thought you knew about Quantum Physics is Different,* 2020, p.65-75.
27. Ford, Kenneth W., *The Quantum World — Quantum Physics for Everyone,* 2005, p.196.
28. Becker, Adam, *What is Real — The unfinished quest for the meaning of quantum physics,* 2019, p.100.
29. Smolin, Lee, *Einstein's Unfinished Revolution — The Search for What Lies Beyond the Quantum,* 2019, p.100.
30. Becker, Adam, *What is Real — The unfinished quest for the meaning of quantum physics,* 2019, p.89.
31. James, Tim, *Fundamental — How Quantum and Particle Physics Explain Absolutely Everything*, 2020, p.108. James in pointing out that de Broglie admitted he made a mistake introducing wave-particle duality and that electrons and protons were particles only. He goes on to say, 'They do not have wave character, but instead were surrounded by some background substance that did have waves in it. The particles were pushed around by these invisible "guide waves" and thus would appear to move in wave-like trajectories.' According to the Tor Model, de Broglie got it right and that the background substance is Penergy. Unfortunately, no one believed him.
32. Becker, Adam, *What is Real — The unfinished quest for the meaning of quantum physics,* 2019, p.102.

33. Baggott, Jim, *Quantum Reality — The Quest for the Real Meaning of Quantum Mechanics — A Game of Theories,* 2020, p.181.
34. Ball, Philip, *Beyond Weird — Why Everything you Thought you knew about Quantum Physics is Different,* 2020, p.110.
35. Becker, Adam, *What is Real — The unfinished quest for the meaning of quantum physics,* 2019, p.150.
36. Becker, Adam, *What is Real — The unfinished quest for the meaning of quantum physics,* 2019, p.150.
37. Becker, Adam, *What is Real — The unfinished quest for the meaning of quantum physics,* 2019, p.221.
38. Clegg, Brian, *The God Effect — Quantum Entanglement, Science's Strangest Phenomenon,* 2006, p.34.
39. Becker, Adam, *What is Real — The unfinished quest for the meaning of quantum physics,* 2019, p.233.
40. Nomura, Yasunori & Bill Poirier & John Terning, *Quantum Physics, Mini Black Holes, and the Multiverse — Debunking Common Misconceptions in Theoretical Physics,* 2018, p.22.
41. Becker, Adam, *What is Real — The unfinished quest for the meaning of quantum physics,* 2019, p.16.
42. Smolin, Lee, *Einstein's Unfinished Revolution — The Search for What Lies Beyond the Quantum,* 2019, p.128.
43. Nomura, Yasunori & Bill Poirier & John Terning, *Quantum Physics, Mini Black Holes, and the Multiverse — Debunking Common Misconceptions in Theoretical Physics,* 2018, p.64.
44. Nomura, Yasunori & Bill Poirier & John Terning, *Quantum Physics, Mini Black Holes, and the Multiverse — Debunking Common Misconceptions in Theoretical Physics,* 2018, p.38.
45. Carroll, Sean, *Something Deeply Hidden — Quantum Worlds and the Emergence of Spacetime,* 2020, p.20.
46. Ball, Philip, *Beyond Weird — Why Everything you Thought you knew about Quantum Physics is Different,* 2020, p.114.
47. Becker, Adam, *What is Real — The unfinished quest for the meaning of quantum physics,* 2019, p.17.

48. Carroll, Sean, *Something Deeply Hidden — Quantum Worlds and the Emergence of Spacetime,* 2020, p.18.

49. Nomura, Yasunori & Bill Poirier & John Terning, *Quantum Physics, Mini Black Holes, and the Multiverse — Debunking Common Misconceptions in Theoretical Physics,* 2018, p.63. Dr. Nomura says, '...Copenhagen [interpretation] does make certain metaphysical claims. One of these is that wavefunction collapse occurs as a result of measurement. Another is that physical systems do not actually *possess* attributes until those attributes are measured.' Her point is that neither of those positions are backed by scientific evidence.

50. Kaku, Michio, *Visions- How Science will Revolutionize the 21st Century,* 1997, p.351.

51. Becker, Adam, *What is Real — The unfinished quest for the meaning of quantum physics,* 2019, p.97.

52. Rovelli, Carlo, *Reality is Not What it Seems — The Journey to Quantum Gravity,* 2017, p.178. Dr. Rovelli devotes an entire chapter to time, entitled *Time Does Not Exist.*

53. Becker, Adam, *What is Real — The unfinished quest for the meaning of quantum physics,* 2019, p.5. Dr. Becker in explaining Bohr's interpretation of quantum mechanics says, 'To them, the theory needs no interpretation, because the things that the theory describes aren't truly real. Indeed, the strangeness of quantum phenomena has led some prominent physicists to state flatly that there is no alternative, that quantum physics proves that small objects simply do not exist is the same objectively real way as the objects in our everyday lives do.'

54. Rovelli, Carlo, *Reality is Not What it Seems — The Journey to Quantum Gravity,* 2017, p.129. Dr. Rovelli says, 'The world is not made up of fields and particles but a single type of entity: the quantum field. There are no longer particles that move in space with the passage of time, but quantum fields whose elementary events happen in spacetime.'

55. Carroll, Sean, *Something Deeply Hidden — Quantum Worlds and the Emergence of Spacetime,* 2020, p.38-40.
56. James, Tim, *Fundamental — How Quantum and Particle Physics Explain Absolutely Everything,* 2020, p.47.
57. Carroll, Sean, *The Big Picture — The Origins of Life, Meaning, and the Universe Itself,* 2017, p.131-132. Dr. Carroll says math and science are completely different endeavors. 'Math is all about proving things, but the things that math proves are not true facts about the actual world. They are implications of various assumptions.'
58. Carroll, Sean, *The Big Picture — The Origins of Life, Meaning, and the Universe Itself,* 2017, p.35. Dr. Carroll says, 'Quantum mechanics has supplanted classical mechanics as the best way to talk about the universe at a deep level. Unfortunately, and to the chagrin of physicists everywhere, we don't fully understand what the theory actually is.'
59. Smolin, Lee, *Einstein's Unfinished Revolution — The Search for What Lies Beyond the Quantum,* 2019, p.29.
60. Ford, Kenneth W., *The Quantum World — Quantum Physics for Everyone,* 2005, p.128.
61. Nomura, Yasunori & Bill Poirier & John Terning, *Quantum Physics, Mini Black Holes, and the Multiverse — Debunking Common Misconceptions in Theoretical Physics,* 2018, p.3.
62. Smolin, Lee, *Einstein's Unfinished Revolution — The Search for What Lies Beyond the Quantum,* 2019, p.25-26.
63. Smolin, Lee, *Einstein's Unfinished Revolution — The Search for What Lies Beyond the Quantum,* 2019, p.137.
64. Smolin, Lee, *Einstein's Unfinished Revolution — The Search for What Lies Beyond the Quantum,* 2019, p.138.
65. Randall, Lisa, *Knocking on Heaven's Door — How Physics and Scientific Thinking Illuminate the Universe and the Modern World,* 2012, p.253.

Chapter Eight

1. Close, Frank, *Antimatter,* 2018, p.29.
2. Hawking, Stephen, *The Theory of Everything,* 2002, p.31
3. Conover, Robert J., *Journey of the Universe — A New Perspective on its Past, Present, and Future Evolution,* 2021.
4. Devereux, Carolyn, *Cosmological Clues — Evidence for the Big Bang, Dark Matter, and Dark Energy,* 2021, p.35.
5. Clark Stuart, *The Unknown Universe — A New Exploration of Time, Space, and Modern Cosmology,* 2016,
6. Kaku, Michio, *The God Equation, The Quest for a Theory of Everything,* 2021, p.173. Dr. Kaku says, 'Any theory of everything will have an infinite number of solutions depending on the initial conditions. But how do you determine the initial conditions of the entire universe? This means you have to input the conditions of the Big Bang from the outside, by hand.' My point in this section is that the more conditions one inputs by hand, the greater the probability of initiating a pathway leading to false theories and realities.
7. Perlov, Delia, and Alex Vilenkin, *Cosmology for the Curious,* 2017, p.301-302.
8. Hawking, Stephen & Leonard Mlodinow, *The Grand Design,* 2010, p.162.
9. Baggott, Jim, *Origins — The Scientific Story of Creation,* 2015, p.4.
10. Krauss, Lawrence M., *A Universe from Nothing — Why There is Something Rather than Nothing,* 2012, p.89.

Bibliography

Alexander, Stephon, *Fear of a Black Universe — An Outsiders Guide to the Future of Physics,* Basic Books, New York, 2021.

Baggott, Jim, *Origins — The Scientific Story of Creation,* Oxford University Press, 2015.

Baggott, Jim, *Quantum Reality — The Quest for the Real Meaning of Quantum Mechanics — A Game of Theories,* Oxford University Press, New York, 2020.

Ball, Philip, *Beyond Weird — Why Everything You Thought About Quantum Physics is Different,* University of Chicago Press, Chicago, IL, 2018

Barnes, Luke A. & Geraint F. Lewis, *The Cosmic Revolutionary's Handbook — Or: How to Beat the Big Bang,* Cambridge University Press, 2020.

Barrow, John D. and Joseph Silk, *The Left Hand of Creation — The Origin and Evolution of the Expanding Universe,* Oxford University Press, New York, 1993.

Becker, Adam, *What is Real — The Unfinished Quest for the Meaning of Quantum Physics,* Basics Books, New York, 2019.

Beckman, Milo, *Math Without Numbers,* Dutton/Penguin Random House LLC, New York, 2021.

Butterworth, Jon, *Atom Land — A Guided Tour Through the Strange (and Impossibly Small) World of Particle Physics,* First Paperback Edition, The Experiment, New York, 2019.

Carroll, Sean, *The Big Picture — The Origins of Life, Meaning, and the Universe Itself,* Dutton, an imprint of Penguin Random House, LLC, New York, Paperback Edition, 2017.

Carroll, Sean, *Something Deeply Hidden — Quantum Worlds and the Emergence of Spacetime,* Dutton, 2019. A good read for understanding Quantum Mechanics, the arguments for a Wave Function, Spacetime, Quantum Field Theory, Entanglement, and Multi-verses.

Carroll, Sean, *Something Deeply Hidden — Quantum Worlds and the Emergence of Spacetime,* Dutton, paperback edition, 2020.

Chaisson, Eric J., *Cosmic Evolution — The Rise of Complexity in Nature,* Harvard University Press, Cambridge, MA, 2001.

Cham, Jorge & Daniel Whiteson, *We Have No Idea — A Guide to the Unknown Universe,* Riverhead Books, New York, Paperback Edition, 2018. An easy read, and a witty but professional look at what we don't know about the universe.

Clark Stuart, *The Unknown Universe — A New Exploration of Time, Space, and Modern Cosmology*, Pegasus Books, New York, 2016.

Clegg, Brian, *Gravity,* Duckworth Overlook, paperback edition, 2016.

Clegg, Brian, *The God Effect — Quantum Entanglement, Science's Strangest Phenomenon,* St. Martin's Press, New York, 2006. An excellent summary of the discoveries and theories behind some of the esoteric aspects of quantum physics.

Cliff, Harry, *How to Make an Apple Pie from Scratch,* Double Day, New York, 2021.

Close, Frank, *Antimatter,* Oxford University Press, United Kingdom, second edition paperback, 2018.

Cowen, Ron, Gravity's Century — From Einstein's Eclipse to Images of Black Holes, Harvard University Press, Cambridge, Massachusetts, 2019.

Davies, Paul, *The Cosmic Blueprint — New Discoveries in Nature's Creative Ability to Order the Universe,* Simon & Schuster, New York, Touchstone Edition, 1989.

Davies, Paul and John Gribbin, *The Matter Myth — Dramatic Discoveries that Challenge our Understanding of Physical Reality*, Simon & Schuster, New York, 1992.

Devereux, Carolyn, *Cosmological Clues — Evidence for the Big Bang, Dark Matter, and Dark Energy*, CRC Press, Taylor & Francis Group, Baca Raton, Florida, 2021

Dine, Michael, *This Way to the Universe,* Dutton/Penguin Random House, LLC, 2022.

Ford, Kenneth W., *The Quantum World — Quantum Physics for Everyone,* Cambridge, Mass.: Harvard University Press, Copyright © 2004, 2005 by the President and Fellows of Harvard College.

Fritzsch, Harald, *The Fundamental Constants — A Mystery of Physics*, World Scientific, Singapore, 2005.

Geach, James, *Five Photons — Remarkable Journeys of Light Across Space and Time,* Reaktion Books, London, 2018

Gott, Richard, *Cosmic Web — Mysterious Architecture of the Universe*, Princeton University Press, Princeton, NJ Paperback Edition, 2018.

Gould, Roy R., *Universe in Creation — A New Understanding of the Big Bang and the Emergence Life,* Harvard University Press, London, 2018.

Greene, Brian, *Until the End of Time — Mind, Matter, and our Search for Meaning in an Evolving Universe,* Alfred A Knopf, New York, 2020.

Greenstein, George, *Symbiotic Universe — Life and Mind in the Cosmos*, William Morrow and Co., New York, 1988.

Guth, Alan H., *The Inflationary Universe — The Quest for a New Theory of Cosmic Origins,* Perseus Books, Reading, Massachusetts, 1997. The story of the reasons for and the development of the *inflation* theory.

Han, M Y, *Quarks and Gluons — A Century of Particle Charges,* World Scientific, Singapore, 1999.

Hawking, Stephen, *Black Holes and Baby Universes, and Other Essays,* Bantam Books, New York, paperback Edition, 1994.

Hawking, Stephen & Leonard Mlodinow, *The Grand Design,* Bantam Books, New York, 2010.

Hawking, Stephen, *A Brief History of Time — From the Big Bang to Black Holes,* Bantam Books, New York, 1990.

Hawking, Stephen, *The Theory of Everything — The Origin and Fate of the Universe,* Jaico Publishing House, 2006.

Impey, Chris, *Einstein's Monsters — The Life and Times of Blackholes,* W.W. Norton & Co., New York, 2019.

Impey, Chris, *How It Began — A Time Traveler's Guide to the Universe*, W.W. Norton & Co., New York, paperback Edition, 2013.

Gregensen, Erik, *Beta-Decay*, Encyclopedia Britannica, www.britannica.com/science/beta-decay, April 21, 2022.

Hossenfelder, Sabine, *Lost in Math — How Beauty Leads Physics Astray*, Basic Books, Paperback Edition, New York, 2020.

Huang, Kerson, *Fundamental Forces of Nature — The Story of Gauge Fields,* World Scientific Publishing Co., Singapore, 2007. A good summary of particle physics and the development of the gauge fields.

James, Tim, *Astronomical -From Quarks to Quasars, the Science of Space at Its Strangest,* Robinson/ Little, Brown Book Group, London, 2020.

James, Tim, *Fundamental — How Quantum and Particle Physics Explain Absolutely Everything*, Pegasus Books, New York, 2020.

Kaku, Michio, *The God Equation — The Quest for a Theory of Everything*, Doubleday, New York, 2021.

Kaku, Michio, *Physics of the Impossible — A Scientific Exploration into the World of Phasers, Force Fields, Teleportation, and Time Travel*, Double Day, New York, 2008.

Kaku, Michio, *Visions- How Science will Revolutionize the 21st Century*, Double Day, New York, 1997.

Kisslinger, Leonard S., *Astrophysics and The Evolution of the Universe*, World Scientific, Singapore, 2017.

Krauss, Lawrence, *The Greatest Story Ever Told — So Far — Why are We Here,* Simon & Schuster, UK, Paperback Edition, 2018.

Krauss, Lawrence M., *A Universe from Nothing — Why There is Something Rather than Nothing,* Free Press — a Division of Simon & Schuster, New York, 2012.

Lindley, David, *The Dream Universe — How Fundamental Physics Lost its Way*, Doubleday, New York, 2020. An easy read recounting the problems with our current theories on how the universe works.

Livio, Mario, *Accelerating Universe — Infinite Expansion, The Cosmological Constant, and the Beauty of the Cosmos*, John Wiley & Sons, Inc., New York and Canada, 2000.

Long, Ariana S., *Ancient Galaxy Clusters Offer Clues About the Early Universe*, Scientific American, Dec. 21, 2021.

McTaggart, Lynne, *The Field — The Quest for the Secret Force of the Universe,* Harper Collins Publishers, New York, 2002. An excellent story of the investigation into the mind, consciousness, and the Zero Point Field.

Munowitz, Michael, *Knowing — The Nature of Physical Law*, Oxford University Press, New York, 2005.

Musser, George, *Spooky Action at a Distance — The Phenomenon that Reimagines Space and Time and What it Means for Black Holes, the Big Bang, and Theories of Everything,* Scientific American/Farrer, Strauss and Giroux, 2015.

Nomura, Yasunori & Bill Poirier & John Terning, *Quantum Physics, Mini Black Holes, and the Multiverse — Debunking Common Misconceptions in Theoretical Physics*, Multiversal Journeys, Springer International, Switzerland, 2018.

Perlov, Delia, and Alex Vilenkin, *Cosmology for the Curious,* Springer International Publishing AG, 2017.

Quinn, Helen R. & Yossi Nir, *The Mystery of the Missing Anti-matter,* Princeton University Press, New Jersey, 2008. A good account of the several arguments regarding why there is only matter in the universe.

Randall, Lisa, *Dark Matter and the Dinosaurs — The Astonishing Interconnectedness of the Universe,* Harper Collins Publishers, New York, 2015.

Randall, Lisa, *Knocking on Heaven's Door — How Physics and Scientific Thinking Illuminate the Universe and the Modern World,* Harper Collins Publishers, New York, paperback edition, 2012.

Rothman, Tony, *A Little Book About the Big Bang,* Harvard University Press, Cambridge, Mass., 2022, p.73.

Rovelli, Carlo, *Reality is Not What it Seems — The Journey to Quantum Gravity,* Riverhead Books, New York, 2017. Good examination of spacetime, reality, relativity, quantum mechanics, and quantum fields.

Schumm, Bruce A., *Deep Down Things — The Breath-Taking Beauty of Particle Physics,* John Hopkins University Press, Baltimore, 2004.

Seife, Charles, *Alpha & Omega — The Search for the Beginning and End of the Universe,* Viking/Penguin Group, New York, 2003.

Smith, Timothy Paul, *Hidden Worlds — Hunting for Quarks in Ordinary Matter,* Princeton University Press, Princeton and Oxford, 2003.

Smolin, Lee, *Einstein's Unfinished Revolution — The Search for What Lies Beyond the Quantum,* Penguin Press, New York, 2019.

Smolin, Lee, *The Trouble with Physics — The Rise of String Theory, the Fall of Science, and What Comes*

Next, Houton Miffin Co., First Mariner Books Edition, New York, 2007.

Sokol, Joshua, *Earliest Black Hole Gives Rare Glimpse of Ancient Universe,* Quanta Magazine, Dec. 6, 2017.

Sutter, Paul S., *Your Place in the Universe — Understanding Our Big Messy Existence,* Prometheus Books, New York, 2018.

Tyson, Neil deGrasse and Donals Goldsmith, *Origins, — Fourteen Billion Years of Cosmic Evolution,* W.W. Norton & Co, New York, 2005, re-issued 2014.

Wilczek, Frank, *A Beautiful Question — Finding Nature's Deep Design,* Penguin Books, 2015.

Wilczek, Frank, *Fundamentals — Ten Keys to Reality,* Penguin Press, New York, 2021.

Index

F

G

H

I

J

K

L

M

N

P

Q

R

S

T

U

V

W

About the Author

Robert J. Conover is a Vietnam era army, helicopter pilot who after graduating from college with a degree in philosophy took up studying particle physics and cosmology with a passion. Intrigued by the new discoveries taking place in particle physics and their ramifications on our picture of the early universe, he was drawn into finding answers to the many questions, conflicts, and unknowns that arose from those discoveries.

A New Vision of the Early Universe is a culmination of many years of research and study of the writings of many particle physicists, cosmologists, and mathematicians. The depth of Bob's research, his attention to detail, and his philosophical approach to problem solving make him the ideal person to write this book.

Bob has retired from a career as a civil investigator and mediation negotiator, during which time he authored *Strategic Tort-Mediation Negotiation.* Bob resides in San Luis Obispo, California, where he continues to read, write, create board games, and cultivate the minds of his five grandchildren.

Made in the USA
Las Vegas, NV
30 August 2023

76819337R00144